BIOCHEMISTRY RESEARCH TRENDS

CYSTEINE

SOURCES, USES AND HEALTH EFFECTS

BIOCHEMISTRY RESEARCH TRENDS

Additional books and e-books in this series can be found on Nova's website under the Series tab.

BIOCHEMISTRY RESEARCH TRENDS

CYSTEINE

SOURCES, USES AND HEALTH EFFECTS

TARAN SAUNDERS
EDITOR

Copyright © 2021 by Nova Science Publishers, Inc.

All rights reserved. No part of this book may be reproduced, stored in a retrieval system or transmitted in any form or by any means: electronic, electrostatic, magnetic, tape, mechanical photocopying, recording or otherwise without the written permission of the Publisher.

We have partnered with Copyright Clearance Center to make it easy for you to obtain permissions to reuse content from this publication. Simply navigate to this publication's page on Nova's website and locate the "Get Permission" button below the title description. This button is linked directly to the title's permission page on copyright.com. Alternatively, you can visit copyright.com and search by title, ISBN, or ISSN.

For further questions about using the service on copyright.com, please contact:
Copyright Clearance Center
Phone: +1-(978) 750-8400   Fax: +1-(978) 750-4470   E-mail: info@copyright.com.

NOTICE TO THE READER

The Publisher has taken reasonable care in the preparation of this book, but makes no expressed or implied warranty of any kind and assumes no responsibility for any errors or omissions. No liability is assumed for incidental or consequential damages in connection with or arising out of information contained in this book. The Publisher shall not be liable for any special, consequential, or exemplary damages resulting, in whole or in part, from the readers' use of, or reliance upon, this material. Any parts of this book based on government reports are so indicated and copyright is claimed for those parts to the extent applicable to compilations of such works.

Independent verification should be sought for any data, advice or recommendations contained in this book. In addition, no responsibility is assumed by the Publisher for any injury and/or damage to persons or property arising from any methods, products, instructions, ideas or otherwise contained in this publication.

This publication is designed to provide accurate and authoritative information with regard to the subject matter covered herein. It is sold with the clear understanding that the Publisher is not engaged in rendering legal or any other professional services. If legal or any other expert assistance is required, the services of a competent person should be sought. FROM A DECLARATION OF PARTICIPANTS JOINTLY ADOPTED BY A COMMITTEE OF THE AMERICAN BAR ASSOCIATION AND A COMMITTEE OF PUBLISHERS.

Additional color graphics may be available in the e-book version of this book.

Library of Congress Cataloging-in-Publication Data

ISBN: 978-1-53619-033-5

Published by Nova Science Publishers, Inc. † New York

Contents

Preface		vii
Chapter 1	Applications of Cysteine in Health and Industries *Anupama R. Prasad, Mathew Kuruvilla and Abraham Joseph*	1
Chapter 2	Therapeutic Use of N-Acetylcisteine for Obsessive Compulsive Disorder: A New Avenue *Flavia di Michele*	31
Chapter 3	Comparative Studies of Cysteine Release from Different Mesoporous Silica: MCM-41 and MCM-48 *Priyanka D. Solanki and Anjali U. Patela*	57
Chapter 4	Optical Sensors for Cysteine: A Brief Overview *Goutam K. Patra, Amit K. Manna and Meman Sahu*	85
Index		103

PREFACE

Cysteine is a naturally occurring, sulfur-containing amino acid with a thiol group and is found in most proteins. Cysteine is a common constituent in health-care products like protein powders, body-building shakers, and smoothies. In medicine, it is used for the treatment of diabetes, inflammation, angina, chronic bronchitis, cardiovascular disease, flu and osteoarthritis. Also, some studies suggest that it can enhance lung health in people with chronic obstructive pulmonary disease (COPD). The first chapter of this book focuses on the uses of cysteine in both industries and medicine and health. The second chapter is about the therapeutic uses of N-acetyl cysteine (NAC) for obsessive compulsive disorder. The third chapter describes the functionalization of MCM-41 and MCM-48 by TPA, encapsulation of cysteine into functionalized carriers and their characterization using various physicochemical techniques. The last chapter includes the most recent developments in designing the fluorescent and colorimetric sensors (optical sensors) for selective and sensitive detection of cysteine.

Chapter 1 - Amino acids are compounds of considerable industrial importance particularly in pharmaceuticals and often act as building blocks for drugs, dietary supplements, and nutraceuticals. There are about 20 primary amino acids in proteins that can be classified into essential and non-essential amino acids. They are found either in the free-state or as linear chains in peptides and proteins. Apart from proteins, free amino acids are also being found in biological materials. Cysteine (L-Cysteine) is a semi-essential polar uncharged amino acid that contains thiols (also called mercaptan) which are compounds with sulfhydryl functional group. It is naturally found in the human body and various foodstuffs. L-Cysteine acts as a precursor for the synthesis of glutathione, which is an important antioxidant in the human body. Cysteine is a common constituent in health-care products, bakery products, animal feeds, and cosmetic items. In natural medicine, it is used for the treatment of diabetes, inflammations, Angina, Chronic bronchitis, cardiovascular diseases, flu, and osteoarthritis. Also, some studies suggest that it can enhance lung health in people with chronic obstructive pulmonary disease (COPD). However, many of these health benefits lack clinical evidence and demands extended research. Amino acids were identified as green corrosion inhibitors for commercial metals owing to their easy availability at low cost and a high degree of purity. The researches have revealed the corrosion inhibitor action of cysteine for commercial metals in acidic environments. Herein, the chapter propose to deliver the health and industrial potentials of cysteine encountered so far.

Chapter 2 - Obsessive-compulsive disorder (OCD) is often a lifetime and disabling disorder since standard therapies both pharmacological and psychosocial may not be efficacious. A dysfunction of glutamatergic neurotransmission and neuroimmune abnormalities might be involved in the pathophysiology of this disorder. Lately, preclinical and clinical studies have shown a neuroprotective role for N-acetyl cysteine (NAC) by acting in two ways: as a modulator of inflammatory pathways and via the modulation of synaptic release of glutamate in cortico-subcortical brain regions. This chapter explores the therapeutic use of NAC in augmentation among individuals with refractory- OCD to the first- line pharmacological interventions, or in monotherapy, reviewing the published clinical studies. The possible benefit mechanisms of NAC for this disorder will be discussed, given its specific property of stimulating the formation of glutathione in the brain. Nutraceutical supplementation with NAC in OCD treatment may be important not only to improve psychiatric symptomatology, but also for its optimal safety and tolerability profile. This is of great interest especially considering the need for treatment special populations affected by OCD, such as children, youngsters and elders. Finally, the nutraceutical approach represents a good choice from a psychological perspective, given its better acceptance by the patients compared to pharmacological treatment.

Chapter 3 - The present chapter describes the functionalization of MCM-41 and MCM-48 by an inorganic moiety, 12-tungstophophoric acid (TPA), Cysteine loading, characterization and *in vitro* release of Cysteine at body temperature under different conditions. A study on Release kinetics was carried out using First order release kinetic model while the mechanism was by Higuchi model. Further, to see the influence of TPA on release rate, release profile obtained from pure MCM-41 and MCM-48 were compared with functionalized materials i.e., TPA-MCM-41 and TPA-MCM-48. Finally, all data were correlated with geometry of the supports.

Chapter 4 - Amino acids are crucially involved in an innumerable of biological processes. Any irregular changes in physiological level of amino acids often manifest in common metabolic disorders, serious neurological conditions and cardiovascular diseases. Among the amino acid series cysteine plays a major role in various physiological processes like protein synthesis, detoxification and metabolism of living organism. Its deficiency can cause several problems like hematopoiesis disease, retarded growth of children, skin lesion and loss of leucocyte etc. Therefore rapid and selective detection and quantification of cysteine in biological relevant samples has become very essential to its efficient clinical finding in recent years. In this book chapter the authors' objective is to discuss the recent developments in designing the fluorescent and colorimetric sensors (optical sensors) for selective and sensitive detection of cysteine.

In: Cysteine: Sources, Uses and Health Effects   ISBN: 978-1-53619-033-5
Editor: Taran Saunders   © 2021 Nova Science Publishers, Inc.

Chapter 1

APPLICATIONS OF CYSTEINE IN HEALTH AND INDUSTRIES

*Anupama R. Prasad[1], Mathew Kuruvilla[2] and Abraham Joseph[1],**

[1]Department of Chemistry, University of Calicut,
Calicut University P O, Kerala, India
[2]Department of Chemistry, St. Thomas College, Ranni,
Patahanamthitta, Kerala, India

ABSTRACT

Amino acids are compounds of considerable industrial importance particularly in pharmaceuticals and often act as building blocks for drugs, dietary supplements, and nutraceuticals. There are about 20 primary amino acids in proteins that can be classified into essential and non-essential amino acids. They are found either in the free-state or as linear chains in peptides and proteins. Apart from proteins, free amino acids are also being found in biological materials. Cysteine (L-Cysteine) is a semi-essential polar uncharged amino acid that contains thiols (also called mercaptan) which are compounds with sulfhydryl functional group. It is naturally

---

* Corresponding Author's E-mail: drabrahamj@gmail.com.

found in the human body and various foodstuffs. L-Cysteine acts as a precursor for the synthesis of glutathione, which is an important antioxidant in the human body. Cysteine is a common constituent in healthcare products, bakery products, animal feeds, and cosmetic items. In natural medicine, it is used for the treatment of diabetes, inflammations, Angina, Chronic bronchitis, cardiovascular diseases, flu, and osteoarthritis. Also, some studies suggest that it can enhance lung health in people with chronic obstructive pulmonary disease (COPD). However, many of these health benefits lack clinical evidence and demands extended research. Amino acids were identified as green corrosion inhibitors for commercial metals owing to their easy availability at low cost and a high degree of purity. The researches have revealed the corrosion inhibitor action of cysteine for commercial metals in acidic environments. Herein, the chapter propose to deliver the health and industrial potentials of cysteine encountered so far.

1. INTRODUCTION

L-cysteine (Cysteine) is a semi-essential amino acid being traded as a dietary supplement. It has a natural occurrence in the human body and foods including meat, dairy products, eggs, nuts, seeds, and legumes [1, 2]. According to the Kyte and Doolittle scale cysteine is a polar amino acid with a positive hydropathy index. Usually, it is not found at the surface of proteins even after ionization. The multiple disulfide bridges in proteins can resist the thermal-denaturation and the biological activity will be preserved at rigorous conditions. Cysteine has a key role in stabilizing the three-dimensional structure of the protein which is extremely important for the extracellular proteins possibly be exposed to harsh bio-environments. Disulfide bridges are generated when the cysteine moiety gets oxidized and stabilizes the ternary and quaternary structure of proteins. Since the sulfhydryl side chain can strongly bind the metal atoms, many of the proteins employ cysteine to hold the metal atoms in position. In the human body, cysteine is found in beta-keratin which is an important constituent in the skin, hair, nails, etc. The existence of a large number of disulfide bonds results in the hard and flexible keratin, whereas a small number result in smooth keratin. Also, the human body uses cysteine to produce glutathione (an antioxidant) and amino acid taurine. The human body also makes use of cysteine as an energy source by converting into glucose, for enzyme catalysis, transcriptional regulation, and protein folding [3-8].

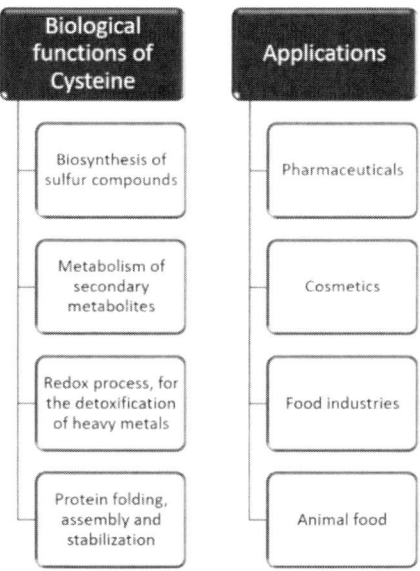

Figure 1. Biological functions and applications of cysteine [4].

Cysteine is a common constituent in health-care products like protein powders, body-building shakes, and smoothies. In natural medicine, it is used for the treatment of diabetes, inflammations, Angina, Chronic bronchitis, cardiovascular diseases, flu, and osteoarthritis. Also, some studies suggest that it can enhance lung health in people with chronic obstructive pulmonary disease (COPD) [9]. Many of these health benefits demands extended research. The health benefits including the effectiveness in the treatment for diabetic colitis, GHS deficiency, oral cavity infections, genetic defects, metabolic disorders, and the overproduction of free radicals have been validated by the research community. Also, cysteine can be used as a green corrosion inhibitor for commercial metals in acidic environments. On the other hand, cysteine has been widely employed in the food and cosmetic industries. Food industries make use of cysteine in bakery applications and animal feeds. In cosmetics, it is used for permanent hair-wave production, nail care, anti-aging, and anti-atrophy skincare products.

2. CYSTEINE FOR HEALTH CARE

Cysteine is semi-essential amino acid since our body can produce it from methionine and serine. The synthesis of cysteine in our body takes place via transmethylation reaction which produces homocysteine as a product which is further converted into cysteine by the transsulfuration reaction [10-12]. Once the nutritional consumption of these two amino acids is deficient cysteine becomes essential. The foods of high protein content like chicken, cheese, eggs, legumes, yogurt, etc. contains sufficient cysteine content.

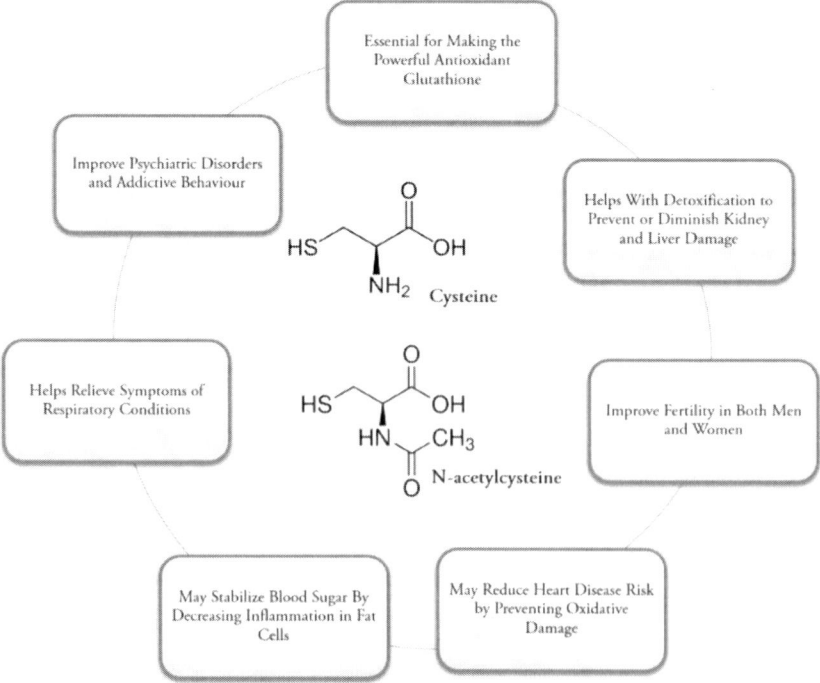

Figure 2. Health benefits of cysteine and NAC supplement.

Applications of Cysteine in Health and Industries

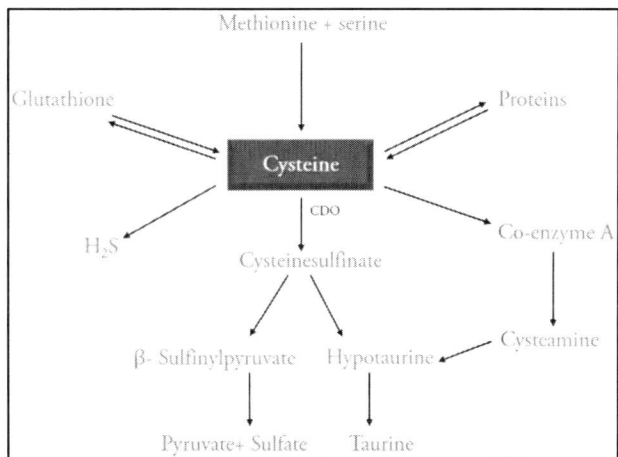

Figure 3. Summary of cysteine metabolism [15].

N-acetyl cysteine (NAC), a precursor of L-cysteine is a medical supplement which is usually recommended for deficiency. It is essential to maintain adequate cysteine levels via proper diet or to have the supplements to restock the powerful anti-oxidants glutathione and taurine which promote good brain health, respiratory conditions, and fertility. Figure 2 details the health benefits of cysteine and NAC supplements. Cysteine is also required for the production of collagen which affects the skin tone. The amino acid supplements are available as single and in combinations as multi-vitamins, proteins, and other food supplements. We have these supplements in powder, tablet, or fluid forms. However, the regular intake of a single amino acid supplement for the long term may result in nitrogen imbalance, a negative impact on kidney function, and growth problems in children. Also, it is not advisable in time of pregnancy, breastfeeding, and for diabetic, cystinuria patients.

There is a substantial increase in research interest on cysteine for the last few decades associated with the nutraceutical industries and personalized medicine. The use of natural products in potential areas including the development of medicines, pharmaceuticals, cosmetics, and benign therapies is currently a trend worldwide. The research on natural products and their incorporation to address several challenging global issues had a momentous expansion. A large number of drugs and supplements were developed and commercialized for the treatment of several diseases [13, 14]. As the structure reveals, cysteine has a thiol group at the end of the side chain which accounts for the high reactivity and biological roles in the human body. Figure 3 represents the extract of cysteine metabolism in the human body.

3. Cysteine Pharmaceuticals and Clinical Trials

L-cysteine is currently being used for the development of several drugs and a concipient of many formulations [16]. However, there is still controversy regarding the impact of cysteine in human health when used in drugs and intake as a dietary supplement because of the lack of evidence-based clinical trials. Flourishing of personalized medicine of natural origin was accelerated by the nutritional habits in developed countries, particularly concerning the importance of nutrition to keep and re-establish good health status. Moreover, dissatisfaction with the traditional treatment which was incapable of curing or release some symptoms. Also, the chemically synthesized drugs were getting refused by a considerable global population.

As we mentioned in the previous session, the research on cysteine in various aspects has received great interest, mainly in the last four decades and there was a significant increase in scientific publications. The analysis reveals that 92% of these publications were based on the physicochemical properties and biological activity of L-cysteine. Review and clinical trials hold the remaining percent 6.45 and 1.29 respectively [3]. Based on the clinical trials, recent research publications recognize the possible advantages and disadvantages of using amino acids for medicines and nutritional supplements. In this context, the use of L-cysteine and its combinations with other natural compounds was identified to have both positive and negative results. Tables 1and 2 have been accessed from a review report published in 2018 by Noelia et al., which summarizes the effect of L-cysteine intake (alone and in combination with other molecules like amino acids, vitamins, etc.,) as nutritional supplement and medicines.

Table 1. Summary of the potential effects of L-Cys (alone) supported by studies based on clinical trials [3]

Examples of L-Cysteine Usage	Effects	Ref.
Nutritional therapy in children with severe edematous malnutrition	Restoration of the rate of synthesis and the concentration of glutathione during the first phase of treatment	[17]
Scarring of the cornea after a photoreactive keratectomy	Reduced average time of scarring	[18]
Nutritional therapy in Ictus patients	Reduced risk of cardiovascular accident	[19]
Hair care	Reduced hair loss and increased hair strengthening abilities	[20]
Protection of digestive system	Reduction in the concentration of acetaldehyde by avoiding exposure in cases of achlorhydria	[21]
Treatment chronic inflammation	Increased antioxidant status	[22]
Prevention of upper digestive tract cancer and breast cancer	Decrease of acetaldehyde in saliva or it can be used as part of metabolic starvation therapy	[23, 24]
Indicator for the control of cardiovascular diseases	Pro-inflammatory signalling	[25]
Treatment of erythropoietic porphyria	Photosensitivity improvement	[26]
Treatment of type-2 diabetes	Control of glycaemia and vascular inflammation	[27-29]

In summary, the published clinical trial reports recommend L-Cysteine comprised medicines have the health benefits: antioxidant power, regulation of the mucolytic function, strengthening of the hair, improvement of the functions of the immune system, protection, and detoxification of the liver, promotion or elimination of heavy metals, prevention of heart disease, diabetes prevention, delay of aging, and the protection of the digestive system. Also, protuberant clinical trials were performed to analyze the effect of Cysteine supplement NAC [35, 36]. The supplement promotes the production of glutathione and owing to the anti-oxidant property it is medicated for the treatment of various ailments associated with the generation of free radicals in our body. NAC is also identified as a potential drug that promote the stubborn mucous discharge. The research outcomes based on the clinical trials have been listed in Table 3.

Table 2. Summary of the potential effects of L-Cys (combined with other compounds) supported by studies based on clinical trials [3]

Composition of the Mixture	Examples of Usage	Effects	Ref.
L-Cysteine + Glycine	Nutritional therapy in elderly HIV + patients Treatment of oxidative stress during aging	Improved oxidation of carbohydrates, insulin sensitivity and body composition. Increased synthesis of glutathione and decreases oxidative stress levels	[30, 31]
L-Cysteine + glycine + dithreonine	Treatment of hypostatic ulcer	Reduced pain and improved degree of ulcer healing.	[32]
L-Cysteine + vitamin D	Treatment of patients with type-2 diabetes	Increased levels of glutathione and decreased levels of triglycerides.	[29]
L-Cysteine + basic fibroblast growth factor (bFCF)	Treatment of corneal epithelium after photoreactive keratectomy in patients affected by myopia	Reduced time of resurfacing corneal.	[33]
L-Cysteine + theanine	Improvement of well-trained athletes' performance	Restoration of the attenuation of the activity of Natural Killer cells.	[34]

Table 3. Summary of the potential effects of NAC supported by studies based on clinical trials [3]

Examples of usage	Effects	Ref.
Treatment of methamphetamine-dependent patients	Methamphetamine dependence disease	[37]
Performance of athletes undergoing strenuous physical training	Redox equilibrium and adaptation processes improve	[38]
Treatment of Thalassemia	Oxidative stress and DNA damage decrease	[39, 40]
Protection against the carcinogenic effect of tobacco	Modulation of biomarkers associated with cancer	[41]
Treatment of bacterial meningitis	Antioxidant role	[42, 43]
Treatment against influenza virus	Proliferation of the virus is inhibited	[43, 44]
Mucolytic expectorant and treatment of respiratory tract infections	The viscosity decreases and facilitates the removal of mucus	[45]
Specific antidote for acetaminophen overdose	Regeneration of glutathione levels	[43, 46-48]
Cardiovascular complications in patients with diabetes Treatment of type-2 diabetes	Attenuation of cardiovascular complication	[49-51]
Prevention of cardiovascular diseases	Reduction of plasma concentrations and homocysteine levels	[52]
Treatment of chronic hepatitis C	Increase in glutathione and improvement in response to treatment with interferon	[53, 54]
Treatment of patients with acute liver failure	Reduced IL-17 levels	[55-58]
Treatment of nephropathic cystinosis	Reduced oxidative stress and improved renal function	[59]
Treatment of noise-induced hearing loss	Protective effect. Hearing loss is reduced	[60]
Treatment of cocaine addiction	It acts as an anti-relapse agent in abstinent subjects	[61]

4. RESEARCH OUTCOMES IN BIOMEDICAL APPLICATIONS

The low molecular mass thiol and their disulfides of L-Cysteine are critical cellular components that play numerous important roles. The reduced form of glutathione (L-Cysteine is the precursor) plays a fundamental role in the defense of the organism against damage caused by oxidative stress [62]. The compound neutralizes reactive particles that can cause damage to cells and tissues. Thus, diet supplementation with L-Cysteine restores the synthesis of glutathione in cases in which it has been compromised, thus improving the redox balance and promoting the reduction of oxidative stress. Also, the elimination of free radicals may be associated with certain benefits, such as in the case of reduced healing time following certain surgical procedures- photorefractive keratectomy, for instance [63]. Besides, the antioxidant role of L-Cysteine is also related to a reduced risk of a cerebrovascular accident and reduced noise-induced hearing loss [64, 65].

Excessive consumption of alcohol is believed to be a risk factor for the development of cancer in the upper gastrointestinal tract, due to exposure to acetaldehyde, which is carcinogenic to humans. L-Cysteine intake reduces the concentration of acetaldehyde in saliva, thereby decreasing the exposure of the gastrointestinal tract to this compound and subsequently the risk of cancer [66] and also reduces the risk of stroke. The consumption of certain amino acids, among which L-Cysteine, may give advantages to certain cardiovascular benefits, such as reduced arterial stiffness or reduced blood pressure, thereby fighting some risk factors related to vascular accidents in healthy women [67].

Polyethylene terephthalate (PET) is a linear, aromatic polyester that has been shown to have the desired mechanical strength and durability for various biomedical applications [78]. PET is an inert polymer and lacks active functional groups on the surface to attach biologically active molecules. It has been used in a wide range of medical applications, including vascular prostheses, heart valve sewing cuffs, implantable sutures, and surgical mesh [69-71]. However, PET applications involving the blood, hemocompatibility is a major concern and in this case surface modification of the PET offers an imperative approach to improve hemocompatibility. Hence, surface functionalization of PET would aid in immobilizing biomolecules which could potentially improve hemocompatibility. Recently, various techniques, including hydrolysis, reduction, glycolysis, aminolysis, and amination [72-76], are used to introduce reactive functional groups on PET surfaces for obtaining a successful outcome in hemocompatibility. Further while assessing the effect of PET modification on surface and bulk properties, it is also critical to analyze whether biologically active molecules can be attached to the modified polymer. PET is used in many blood-contacting systems in which PET as a foreign material activates platelets and improves the platelet aggregation. It has been reported that Cysteine has enabled to improve hemocompatibility when immobilized to aminolysed PET. In such cases, Cysteine utilizes endogenous nitric oxide (NO) to inhibit platelet activation and aggregation. A rapid transnitrosation reaction also occurs between S-nitroso proteins (primarily S-nitroso albumin) and cysteine [77]. However, S-nitroso cysteine formed became unstable and nitric oxide is also released. The released NO acts as a potent inhibitor of platelet activation and aggregation.

Poly Amino Acids (PAA) in general have gained substantial attention from the researchers as biocompatible polymers due to various reasons such as their degradation products are biocompatible L-amino acids and (ii) the absolute isotacticity properties of the biopolymers. The secondary structures formed out of isotacticity of the polymer contribute to the overall biocompatibility of the biopolymers employed. The stable secondary structures of PAAs adopt in aqueous solution exhibits superior behaviour over conventional synthetic polymers and it promotes self-assembly which can be exploited in biomedical applications, such as drug delivery, tissue engineering, and polymer coatings [78]. Poly(l-Cysteine), forms preferably β-sheets [79], and aside from the secondary structure, the amphiphilicity and reactivity of PAAs imparted by their side chains to determine the physical properties and subsequent utilization in biomedicine. The redox-active thiol group in p(l-Cys) oxidizes readily thereby cross-linking the polymer. The biocompatibility of PAAs could be improved by developing block copolymers with other biocompatible polymers including PEG and poly(oxazoline). PEG occupies a pivotal role among the biocompatible polymers because it is polyether widely used as a biomedical polymer, capable of inducing a stealth character upon PEGylated moieties and surfaces [80]. These copolymers exhibit low toxicity and high hydrophilicity [81], which makes them acceptable for therapeutic applications, such as drug and gene delivery.

Hyaluronic acid (HA) is a non-sulfated anionic linear polysaccharide consisting of b-(1, 4) linked D-glucuronic acid-b-(1, 3)- N-acetyl-D-glucosamine disaccharide repeats. It is ubiquitous in the human body, particularly in the synovia of joints, the corpus vitreum of the eyes, and the dermis of the skin [82]. As an essential signaling biomacromolecule, HA is located in the extracellular matrix (ECM) both at the cell surface and inside the cell. HA has a short half-life of only about 1 to 2 days in tissue, limiting its applications as a biocompatible, biodegradable, and non-immunogenic polymer [83]. Chemical modification and crosslinking of HA pave the way to achieve longer residence time *in vivo* while maintaining biocompatibility and viscoelastic properties of the naturally occurring biomolecule [84]. A new strategy for the preparation of Cys–HA useful for *in situ* formations of hydrogels by native chemical ligation has been developed. This method preserves the free carboxylic acids by conjugating the cysteine through a stable ether linkage via the HA hydroxyl group.

Amino acids (AA) are one of the most important biomolecules, possessing different functional groups such as thiol (-SH), carboxyl (-COOH), an amino ($-NH_2$) groups, which allow their adsorption on inorganic surfaces and thus it is used as the surface modifier of inorganic materials [85]. Kotok's group graft L-cysteine on the surface of the graphene quantum dot and prepared chiral graphene quantum dots [86]. On the contrary, Wei's et al. used L-cysteine as chiral ligand and prepared chiral No nanoparticles [87]. However, a few reports are available about the *in vivo* biosafety of amino acid modified No. L-cysteine, as a model amino acid, was grafted on the surface of ZnO nanoparticles and formed a kind of biosafety ZnO hybrid (ZnO@PDA-g-L-Cys) via mussel inspired method [88].

5. APPLICATIONS IN COSMETICS AND FOOD INDUSTRIES

Amino acids and their salts are extensively used as ingredients in the cosmetic industry. The function of amino acids is targeted on skin and hair conditioning. They also function as an oral care agent, antioxidants, hair waving/straightening agents, fragrant ingredients, and buffering agents. Cysteine and cystine are related as thiol (reduced monomer) and the disulfide (oxidized dimer) forms of the same structure. These two molecules play major roles in reversible cellular redox chemistry and can serve a similar function in hairdressings. On the cosmetics front, cysteine is used widely as a substitute for thioglycolic acid in permanent hair wave preparation due to its ability to break disulfide bonds in keratin. Thioglycolic was used earlier to prepare hair for perms but it has an unpleasant door and is an allergenic [89]. Further, cystine (two cysteine molecules formed cystine) is used in nail care as it promotes proper fingernail growth, hardness, and functionality [90]. The acetylated cysteine (*N*-acetylcysteine) is being used in the formulation of safe and effective products for antiaging and anti-atrophy skincare products [91].

Hair blends fortified with L-Cysteine help to strengthen hair. Keratin is one of the most abundant proteins in the skin and the hair and contains high amounts of L-Cysteine as building blocks. L-Cystine forms disulfide bridges, which provide strength and rigidity to keratin. Further, the use of blends fortified with L-Cys promotes the repair of structural lesions and slows down hair loss experienced by patients affected by certain disorders (diffuse alopecia), for instance [92]. We are familiar with the keratin treatment for healthy hair. Cysteine is naturally present in keratin and affects strength, smoothness, and flexibility. Cysteine hair treatment was introduced very recently and slowly being popularized which is used for hair straightening and smoothing. The treatment uses a cysteine complex very similar to the keratin which improves dull hair, clear frizz, and dryness. Both keratin and cysteine treatments are temporary and expensive yet the cysteine treatment claims some good features which make difference. It doesn't use chemicals, thus safe for children above 12 years and pregnant women. Keratin treatment commonly uses formaldehyde. Cysteine treatment can be done from home and is milder treatment as compared to keratin. Moreover, the treated hair looks natural and even the newly grown hair doesn't make much difference from the treated one.

Amino acids generally found in dietary protein, are released as the protein digested in the gastrointestinal tracts. Many of the amino acids are considered to be essential and may be obtained through the diet. In the food industry, particularly in bakery applications, cysteine is used as flour additives to break up the gluten in the flour, thus reducing its stickiness and facilitating the kneading of the dough. Theoretically, protein-protein interactions through covalent disulfide bond formation can be disrupted by adding cysteine which would increase the elasticity of the formed dough, helping it to rise during baking [93]. Another important aspect to point out is the potential use of L-Cysteine in processes related to food conservation and processing. L-Cysteine shows antioxidant, chelating, and flavouring properties which could be useful in food industries [94]. As animal feed, cystine is considered efficacious in partially meeting the requirements of sulfur-containing amino acids in all animal species. L-c.

Cysteine has also been reported to have prophylactic potential against nitrate poisoning in ruminant species.

Cysteine proteases are proteins that contain cysteine residues which can catalyze the hydrolysis of peptide, amide, ester, thiol ester, and thiono ester bonds. Proteases are widely used in the baking industry. Protease treatment improves dough relaxing and bread volume, prevents dough shrink back, and allows faster bakery throughput. Cysteine proteases, especially papain and bromelain, are widely used to prepare protein hydrolysates having excellent taste properties because of the absence of bitterness. Proteinases are also applied in the formulation of marinades and tenderizing recipes. Softness and tenderness are the most important factors affecting consumer satisfaction and the perception of taste. Tenderisation can be affected by breaking the cross-links between the fibrous protein of meat (collagen and elastin) or by breaking meat into shreds. Plant cysteine proteases are also used to improve the recovery of protein from slaughterhouse waste and soy processing. The recovered proteins are subsequently used in both the feed and food industries owing to their good nutritional value and excellent functional properties [95-97].

6. CYSTEINE- GREEN CORROSION INHIBITOR

Corrosion is a major global challenge to the petroleum industry and has become a very serious problem with the initiation of common metals used in all walks of life from the beginning. A corrosion inhibitor is a chemical substance that, upon addition to a corrosive environment, results in a reduction of corrosion rate to an acceptable level. Corrosion inhibitors are generally used in small concentrations. The use of inhibitors is one of the most practical methods for protection against corrosion, especially in acid solutions. One of the most efficient and eco-friendly methods widely accepted for the removal of contaminants is the use of low-cost materials [98]. Organic inhibitors that contain nitrogen, oxygen, and sulfur have been widely used in the case of corrosion of copper and its alloys [99]. Considering the highly toxic nature of these organic compounds, amino acids were identified as an alternate, a green corrosion inhibitor with unique chemical properties. The advantages of amino acid inhibitors are that they are available relatively easily at low cost and with a high degree of purity. An array of amino acids has been employed in the corrosion studies of various metals and are listed in Table 4. Cysteine is a very interesting amino acid that contains amino groups [–NH_2], carboxyl group [– COOH], and thiol group [–SH]. It can coordinate with metals through the nitrogen atom, the oxygen atom of the carboxyl group, and the sulfur atom of the thiol group. The amino acid L- Cysteine is used to control the corrosion of mild steel, aluminium and copper. Cysteine has been used to prevent the corrosion of a wide variety of metals.

Table 4. Cysteine as corrosion inhibitor for commercial metals

No.	Metal	Medium	Inhibitor	Ref
1	Copper	Sulphuric acid	L cysteine	100
3	Mild steel	1M HCl	Cysteine, serine amino butyric acid threonine,	101
4	Mild steel	1M HCl	Cysteine Alanine, and S- methyl cysteine	102
5	Cu-Ni alloy	0.5M H_2SO_4	Cysteine	103
6	Copper	Sulphuric acid	Cysteine Schiff base	104
7	Copper	Sulphuric acid	Cysteine and Alanine	105

Studies were carried out to discover the inhibitory effect of sulfur-containing amino acid, L-cysteine, on copper in different concentrations of sulfuric acid (0.5-1.5 M) at different temperatures [100]. Results revealed that L-cysteine does offer an attractive inhibition efficiency. However, with an increase in the concentration of the inhibitor, corrosion rates decreased irrespective of the temperature gradients. This is due to surface adsorption of the inhibitor molecules on the metal which has contributed to a decreased double-layer capacitance and increased polarization resistance. With the increase in the concentration of the medium, the corrosion rate was also enhanced and this is due to the liberation of a high quantum of H+ ions. Based on the results of Tafel polarization studies, it is evident that the amino acid, L-cysteine, could act as a mixed-type inhibitor.

The synergic interaction of amino acids cysteine and alanine with attractive corrosion inhibition efficiency in 1 M sulfuric acid provided significant findings. The inhibition effect of the amino acid cluster advanced with the increased concentration of the inhibitor. However, with the increase in temperature, the inhibition efficiency showed a declining trend. It was attributed that such behaviour of the amino acids on copper in the sulfuric acid medium may be due to surface adsorption of the inhibitor molecules on the metal which contributes to a decrease in the double-layer capacitance and increase in the polarization resistance [106]. Corrosion studies in Cu37Zn brass in neutral and weakly alkaline sulfate solutions were carried out using Cysteine as a green inhibitor. Brass is one of the most widely used copper alloys investigated in depth at different aggressive media at various pH regimes [99]. The inhibitory effect of cysteine is based on the formation of a stable Cu (I)–cysteine complex, and it is more pronounced in a weakly alkaline solution. The amino acid cysteine is recommended, as a green inhibitor, for corrosion protection of brass.

N-acetyl cysteine-based corrosion inhibitor formulations for steel protection in 15% HCl solution was reported [107]. The highlights of the studies are N-acetyl cysteine (NAC) moderately inhibited corrosion of X80 steel. The concentration of the inhibitor and temperature governs the corrosion inhibition process, however, the formulations with other additives enhanced corrosion inhibition effect at the higher temperature. In recent years thrust was also given on the role of amino acids and its Schiff bases as green corrosion inhibitors in preventing corrosion of metals due to its non-toxic nature and biodegradable properties. Schiff bases can be synthesized from relatively inexpensive and simple raw materials. These compounds receive attention because of their potential applications and properties such as anticancer, anticonvulsant, antitumor, antifungal, antibacterial, antitubercular, antioxidant, antimalarial, anti-inflammatory, biological, anti-HIV and pesticidal and corrosion inhibition properties. The role of Schiff base derived from L-cysteine, HDMMA [(E)-2-((2-hydroxyl-4a,8a-dihydronaphthalene-1-yl) methylene amino)-3-mercaptopropanoic acid] as corrosion inhibitor for metallic copper in 1.0 M sulphuric acid was reported [104]. Due to the improved flexibility and diverse structural features, a wide range of Schiff's bases have been synthesized and utilized for extensive applications in medicinal, agricultural, pharmaceuticals, and material science fields. The excellent corrosion inhibition efficiencies offered by different Schiff's bases are yet to be explored. The presence of the electron cloud on the aromatic ring and sp2 hybridized hetero atoms in the molecule enhances the corrosion inhibition activity. The Schiff base HDMMA showed attractive inhibition efficiency even at elevated temperatures.

In the recent past, research efforts oriented in corrosion chemistry are mainly to identify green corrosion inhibitors which are cost effective and capable of reducing the intensity of environmental pollution. In this context, amino acids as well as their Schiff bases are efficient and capable of reducing corrosion level in the metals irrespective of the medium and temperature conditions as reported. However, the efficiency of corrosion inhibition depends on the inhibitor concentration and temperature conditions. The synergism of the amino acids is also effective for corrosion inhibition in metals.

CONCLUSION

L – Cysteine is one of the free amino acids considered to be highly essential and are found in dietary protein. It is widely employed in various areas covering industrial research, healthcare, cosmetic industry, biomedical applications, etc. The amino acid is bestowed with low molecular mass thiol and disulfides act as a precursor for the antioxidant glutathione. The semi-essential amino acid L-Cysteine is synthesized in the human body via transmethylation reaction. It is highly necessary to safeguard adequate cysteine levels through proper diet or through supplements for upholding health status. The potential use of L-Cysteine in processes related to food conservation and processing. The antioxidant, chelating, and flavouring properties of the amino acid are being exploited in the food industries. In recent years research emphasis has been given to highlight the importance of L- Cysteine as a green corrosion inhibitor. The unique chemical properties of the amino acid have the advantage to prevent the corrosion of metals such as mild steel, aluminium, and copper. The amino group [–NH_2], carboxyl group[–COOH], and thiol group[–SH] of the amino acid enable to coordinate with metals through the nitrogen atom, the oxygen atom of the carboxyl group, and sulfur atom of thiol group. Further, the role Schiff base of amino acid L- Cysteine is highlighted as a green corrosion inhibitor. The importance of amino acid L – Cysteine and its salts as ingredients in various industries including healthcare, biomedical, cosmetic, and food industry are highlighted. The research was undertaken on cysteine and its applications in the past few decades. However, further research is warranted to highlight the role of L- Cysteine in modern medicine, nano research, cosmetics, food industry, and as a green corrosion inhibitor independently as well as in unison with other amino acids under various media at altered conditions to tackle several scientific issues.

REFERENCES

[1] Demirkol O, Adams C, Ercal N. Biologically important thiols in various vegetables and fruits. *J Agric Food Chem* 2004; 52: 8151-8154. http://dx.doi.org/10.1021/jf040266f.

[2] Devlin T. M. *Textbook of Biochemistry with Clinical Correlations*, 7th ed.; Chapters 3 and 9; John Wiley & Sons Inc.: Hoboken, NJ, USA, 2010; ISBN 978-0-470-28173-4.

[3] Noelia Clemente P, Manuel Reig G. G, Rosa María M. E. Effects of the Usage of L-Cysteine (L-Cys) on Human Health. *Molecules* 2018; 23: 575-588. 10.3390/molecules23030575.

[4] Nur Izzah I, Yumi Zuhanis H. H, Parveen J, Rashidi O, Hamzah M. S. Production of Cysteine: Approaches, Challenges and Potential Solution. *Int. J. Biotechnol. Wellness Ind*. 2014;3(3): 95-101. ISSN: 1927-3037/14.

[5] Fahey R. C. Biologically important thiol-disulfide reactions and the role of cyst(e)ine in proteins: An evolutionary perspective. *Adv. Exp. Med. Biol*. 1977; 86: 1–30. 10.1007/978-1-4684-3282-4_1.

[6] Bin, P, Huang, R., Zhou. X. Oxidation Resistance of the Sulfur Amino Acids: Methionine and Cysteine. *Biomed. Res. Int*. 2017; 2017: 9584932. 10.1155/2017/9584932.

[7] Go Y. M., Chandler J. D, Jones D. P. The Cysteine Proteome. *Free Radic. Biol. Med*. 2015; 84: 227–245. https://doi.org/10.1016/j.freeradbiomed.2015.03.022.

[8] Meyer A. J, Hell R. Glutathione homeostasis and redox-regulation by sulfhydryl groups. *Photosynth. Res*. 2005; 86: 435–457. 10.1007/s11120-005-8425-1.

[9] *L-Cysteine: Benefits, Side Effects, Dosage, and Interactions*: https://www.verywellhealth.com/the-benefits-of-l-cysteine-89468, accessed on 14/8/2020.

[10] Bin P, Huang R., Zhou X. Oxidation Resistance of the Sulfur Amino Acids: Methionine and Cysteine. *Biomed. Res. Int.* 2017; 2017: 9584932. https://doi.org/10.1155/2017/9584932.

[11] Devlin T. M. *Textbook of Biochemistry with Clinical Correlations*, 7th ed.; Chapters 3 and 9; John Wiley & Sons Inc.: Hoboken, NJ, USA, 2010; ISBN 978-0-470-28173-4.

[12] Stipanuk M. H. Metabolism of sulfur-containing amino acids. *Annu. Rev. Nutr.* 1986; 6: 179–209. https://doi.org/10.1146/annurev.nu.06.070186.001143.

[13] Fontana L, Sáez M. J, Santisteban R, Gil A. Nitrogenus compounds of interest in clinical nutrition. *Nutr. Hosp.* 2006; 21: 14–27. PMID: 16771070.

[14] Eisenberg D. M, Davis R. B, Ettner, S. L, Appel S, Wilkey S, Van Rompay M, Kessler R. C. Trends in alternative medicine use in the United States, 1990–1997: Results of a follow-up national survey. *JAMA* 1998; 280; 1568–1575.

[15] Oja S. S, Saransaari P. Open questions concerning taurine with emphasis on the brain. *Adv. Exp. Med. Biol.* 2015; 803: 409–413. 10.1007/978-3-319-15126-7_31.

[16] Vilanova J. C. Literature review of the subject of a research project. *Radiología* 2012; 54: 108–114. 10.1016/j.rx.2011.05.015.

[17] Badaloo A, Reid M, Forrester T, Heird W.C, Jahoor F. Cysteine supplementation improves the erythrocyte glutathione synthesis rate in children with severe edematous malnutrition. *Am. J. Clin. Nutr.* 2002; 76: 646–652. 10.1093/ajcn/76.3.646.

[18] Meduri A, Grenga P. L, Scorolli L, Ceruti P, Ferreri G. Role of cysteine in corneal wound healing after photorefractive keratectomy. *Ophthalmic Res.* 2009; 41: 76–82. https://doi.org/10.1159/000187623.

[19] Larsson S. C, Hakansson N, Wolk A. Dietary cysteine and other amino acids and stroke incidence in women. *Stroke* 2015; 46: 922–926. https://doi.org/10.1161/STROKEAHA.114.008022.

[20] Petri H, Pierchalla P, Tronnier H. The efficacy of drug therapy in structural lesions of the hair and in diffuse effluvium–comparative

double-blind study. *Schweiz. Rundsch. Med. Prax.* 1990; 79: 1457–1462. PMID: 1709511.

[21] Linderborg K, Marvola T, Marvola M, Salaspuro M, Färkkilä M, Väkeväinen S. Reducing carcinogenic acetaldehyde exposure in the achlorhydric stomach with cysteine. *Alcohol. Clin. Exp. Res.* 2011; 35: 516–522. https://doi.org/10.1111/j.1530-0277.2010.01368.x.

[22] McPherson R. A, Hardy G. Clinical and nutritional benefits of cysteine-enrich protein supplements. *Curr. Opin. Clin. Nutr. Metab. Care* 2011; 14: 562–568. 10.1097/MCO.0b013e32834c1780.

[23] Salaspuro V, Hietala J, Kaihovaara P, Pihlajarinne L, Marvola M, Salaspuro M. Removal of acetaldehyde from saliva by a slow-release buccal tablet of L-cysteine. *Int. J. Cancer* 2002; 97: 361–364. 10.1002/ijc.1620.

[24] Gec, R. C, Toker A. Nonessential amino acid metabolism in breast cancer. *Adv. Biol. Regul.* 2016; 62: 11–17. 10.1016/j.jbior.2016.01.001.

[25] Go Y. M, Jones D. P. Cysteine/cystine redox signalling in cardiovascular disease. *Free Radic. Biol. Med.* 2011; 5: 495–509. 10.1016/j.freeradbiomed.2010.11.029.

[26] Mathews-Roth M. M, Rosner B. Long-term treatment of erythropoietic protopotphyria with cysteine. *Photodermatol. Photoimmunol. Photomed.* 2002; 18:307–309. 10.1034/j.1600-0781. 2002. 02790.x.

[27] Jain S. K. L-cysteine supplementation as an adjuvant therapy for type-2 diabetes. *Can J. Physiol. Pharmacol.* 2012; 90: 1061–1064. https://doi.org/10.1139/y2012-087.

[28] Carter R. N, Morton N. M. Cysteine and hydrogen sulphide in the regulation of metabolism: Insights from genetics and pharmacology. *J. Pathol.* 2016; 238: 321–332. 10.1002/path.4659.

[29] Jain S. K, Micinski D, Huning L, Kahlon G, Bass P. F, Levine S. N. Vitamin d and L-cysteine levels correlate positively with GSH and negatively with insulin resistance levels in the blood of type 2 diabetic patients. *Eur. J. Clin. Nutr.* 2014; 68: 1148–1153. https://dx.doi.org/10.1038%2Fejcn.2014.114.

[30] Nguyen D, Hsu J. W, Jahoor F, Sekhar R. V. Effect of Increasing glutathione with cysteine and glycine supplementation on mitochondrial fuel oxidation, insulin sensitivity, and body composition in older HIV-infected patients. *J. Clin. Endocrinol. Metab.* 2014; 99: 169–177. 10.1210/jc.2013-2376.

[31] Sekhar R. V, Patel S. G, Guthikonda A. P, Reid M, Balasubramanyam A, Taffet G. E, Jahoor F. Deficient synthesis of glutathione underlies oxidative stress in aging and can be corrected by dietary cysteine and glycine supplementation. *Am. J. Clin. Nutr.* 2011; 94: 847–853. https://dx.doi.org/10.3945%2Fajcn.110.003483.

[32] Meduri A, Scorolli L, Scalinci S.Z, Grenga P.L, Lupo S, Rechichi M, Meduri E. Effect of the combination of basic fibroblast growth factor and cysteine on corneal epithelial healing after photorefractive keratectomy in patients affected by myopia. *Indian J. Ophthalmol.* 2014; 62: 424–428. 10.4103/0301-4738.119420.

[33] Kawada S, Kobayashi K, Ohtani M, Fukusaki C. Cystine and theanine supplementation restores high-intensity resistance exercise-induced attenuation of natural killer cell activity in well-trained men. *J. Strength Cond. Res.* 2010; 24: 846–851. 10.1519/JSC.0b013e 3181c7c299.

[34] Cotgreave I. A. N-acetylcysteine: Pharmacological considerations and experimental and clinical applications. *Adv. Pharmacol.* 1997; 38: 205–227. PMID: 8895810.

[35] Samuni Y, Goldstein S, Dean O. M, Berk M. The chemistry and biological activities of N-acetylcysteine. *Biochim. Biophys. Acta 2013*; 1830: 4117–4129. 10.1016/j.bbagen.2013.04.016.

[36] Mousavi S. G, Sharbafchi M. R, Salehi M, Peykanpour M, Karimian Sichani N, Maracy M. The efficacy of N-acetylcysteine in the treatment of methamphetamine depended: A double-blind controlled, crossover study. *Arch. Iran Med.* 2015; 18: 28–33. PMID: 25556383.

[37] Slattery K. M, Dascombe B, Wallace L. K, Bentley D. J, Coutts A. J. Effect of N-acetylcysteine on cycling performance after intensified training. *Med. Sci. Sports Exerc.* 2014; 46: 1114–1123. 10.1249/ MSS.0000000000000222.

[38] Ozdemir Z. C, Koc A, Aycicek A, Kocygit A. N-acetylcysteine supplementation reduces oxidative stress and DNA damage in children with β-thalassemia. *Hemoglobin* 2014; 38: 359–364. 10.3109/03630269.2014.951890.

[39] Rachmilewitz E. A, Weizer-Stern O, Adamsky K, Amariglio N, Rechavi G, Breda L, Rivella S, Cabantchik Z. I. Role of iron in inducing oxidative stress in thalassemia: Can it be prevented by inhibition of absorption and by antioxidants? *Ann. N. Y. Acad. Sci.* 2005; 1054: 118–123. https://doi.org/10.1196/annals.1345.014.

[40] Van Schooten F. J, Besaratina A, De Flora S, D'agostina F, Izzotti A, Camoirano A, Balm A. J, Dallinga J. W, Bast A, Haenen G. R., Laura V Hanen, Paul B, Harumasa S, Nico Van Z. Effects of oral administration of N-acetyl-L-cysteine: A multi-biomarker study in smokers. *Cancer Epidemiol. Biomarkers Prev.* 2002; 11: 167–175. PMID: 11867504.

[41] Klein M, Koedel U, Pfister H. W. N-acetyl-L-cysteine as a therapeutic option in bacterial meningitis. *Der Nervenarzt* 2007; 78: 202–205. 10.1007/s00115-006-2232-6.

[42] Millea P. J. N-acetylcysteine. Multiple clinical applications. *Am. Fam. Physic.* 2009; 80: 265–269. PMID: 19621836.

[43] Uchide N, Toyoda H. Antioxidant therapy as a potential approach to severe influenza-associated complications. *Molecules* 2011; 28: 2032–2052. 10.3390/molecules16032032.

[44] Blasi F, Page C, Rossolini G. M, Pallecchi L, Matera M. G, Rogliani P, Cazzola M. The effect of N-acetylcysteine on biofilms: Implications for the treatment of respiratory tract infections. *Respir. Med.* 2016; 117: 190–197. 10.1016/j.rmed.2016.06.015.

[45] Bass S, Zook N. Intravenous acetylcysteine for indications other than acetaminophen overdose. *Am. J. Health Syst. Pharm.* 2013; 70: 1496–1501. 10.2146/ajhp120645.

[46] Larsen L. C, Fuller S. H. Management of acetaminophen toxicity. *Am. Fam. Phys.* 1996; 53: 185–190. PMID: 8546045.

[47] Blackford M. G, Felter T, Gothard M. D, Reed M. D. Assessment of the clinical use of intravenous and oral N-acetylcysteine in the

treatment of acute acetaminophen poisoning in children: A retrospective review. *Clin. Ther.* 2011; 33: 1322–1330. 10.1016/j.clinthera.2011.08.005.

[48] Xu Y. J, Tappia P. S, Neki N. S, Dhalla N. S. Prevention of diabetes-induced cardiovascular complications upon treatment with antioxidants. *Heart Fail. Rev.* 2014; 19: 113–121. 10.1007/s10741-013-9379-6.

[49] Lasram M. M, Dhouib I. B, Annabi A, El Fazaa S, Gharbi N. A review on the possible molecular mechanism of action of N-acetylcysteine against insulin resistance and type-2 diabetes development. *Clin. Biochem.* 2015; 48: 1200–1208. 10.1016/j. clinbiochem.2015.04.017.

[50] Pereira S, Shah A, Fantus I. G, Joseph J. W, Giacca A. Effect of N-acetyl-L-cysteine on insulin resistance caused by prolonged free fatty acid elevation. *J. Endocrinol.* 2015; 225: 1–7. 10.1530/JOE-14-0676.

[51] Hildebrandt W, Sauer R, Bonaterra G, Dugi K. A, Edler L, Kinscherf R. Oral N-acetylcysteine reduces plasma homocysteine concentrations regardless of lipid or smoking status. *Am. J. Clin. Nutr.* 2015; 102: 1014–1024. https://doi.org/10.3945/ajcn.114. 101964.

[52] Bonkovsky H. L. Therapy of hepatitis C: Other options. *Hepatology* 1997; 26: 143S–151S. 10.1002/hep.510260725.

[53] Czaja A. J. Hepatic inflammation and progressive liver fibrosis in chronic liver disease. *World J. Gastroenterol.* 2014; 20: 2515–2532. https://dx.doi.org/10.3748%2Fwjg.v20.i10.2515.

[54] Stravitz R, Sanyal A. J, Reisch J, Bajaj J. S, Mirshahi F, Cheng J, Lee W. M. Effects of N-acetylcysteine on cytokines in non-acetaminophen acute liver failure: Potential mechanism of improvement in transplant-free survival. *Liver Int.* 2013; 33: 1324–1331.10.1111/liv.12214.

[55] Kemp R, Mole J, Gomez D. Nottingham HPB Surgery Group. Current evidence for the use of N-acetylcysteine following liver resection. *ANZ J. Surg.* 2017; 13: E486-E490. 10.1111/ans.14295.

[56] Grendar J, Ouellet J. F, McKay A, Sutherland F. R, Bathe O. F, Ball C. G, Dixon E. Effect of N-acetylcysteine on liver recovery after

resection: A randomized clinical trial. *J. Surg. Oncol.* 2016; 114: 446–450. 10.1002/jso.24312.
[57] Robinson S. M, Saif R, Sen G, French J. J, Jaques B. C, Charnley R. M, Manas D. M, White S. A. N-acetylcysteine administration does not improve patient outcome after liver resection. *HPB* 2013; 15: 457–462. 10.1111/hpb.12005.
[58] Pache de Faria Guimaraes L, Seguro A. C, Shimizu M. H, Lopes Neri L. A, Sumita N. M, de Bragança A. C, Aparecido Volpini R, Cunha Sanches T. R, Macaferri da Fonseca F. A, Moreira Filho C. A, Maria H. V. N-acetyl-cysteine is associated to renal function improvement in patients with nephropathic cystinosis. *Pediatr. Nephrol.* 2014; 29: 1097–1102. 10.1007/s00467-013-2705-3.
[59] Doost A, Lotfi Y, Moossavi A, Bakhshi E, Talasaz A.H, Hoorzad A. Comparison of the effects of N-acetyl-cysteine and ginseng in prevention of noise induced hearing loss in male textile workers. *Noise Health* 2014; 16: 223–227. 10.4103/1463-1741.137057.
[60] Nocito Echevarria M. A, Andrade Reis T, Ruffo Capatti G, Siciliano Soares V, da Silveira D. X, Fidalgo T. M. N-acetylcysteine for treating cocaine addiction—A systematic review. *Psychiatry Res.* 2017; 251: 197–203. 10.1016/j.psychres.2017.02.024.
[61] Moussawi K, Pacchioni A, Moran M, Olive M. F, Gass J. T, Lavin A, Kalivas P. W. N-acetylcysteine reverses cocaine-induced metaplasticity. *Nat. Neurosci.* 2009; 12: 182–189. 10.1038/nn.2250.
[62] Margaret I. T. Amino acid analysis: an overview. In Cooper C, Packer N, Williams, K. eds., *Amino acid analysis protocols*. Humana Press: New Jersey 2001, pp. 1-7.
[63] Nissen K. E, Stuart B. H, Stevens M. G, Baker A. T. Characterization of aminated poly (ethylene terephthalate) surfaces for biomedical applications. *J Appl Polym Sci* 2008; 107:2394-2403. https://doi.org/10.1002/app.27145.
[64] Vinard, Eloy R, escotes J, Brudon J. R, Guidicelli H, Magne J. L, Patra P, Berruet R, Huc A, Chauchard J. Stability of performances of vascular prostheses retrospective study of 22 cases of human implated

prostheses. *J Biomed Mater Res*, 1988; 22:633-648. 10.1002/jbm.820220705.

[65] Homsy C. A, Mc Donald K. E, Akers W. W, Short C, Freeman B. S. Surgical suture-canine tissue interaction for six common suture types. *J Biomed Mater Res* 1968; 2:215-230. 10.1002/jbm. 820020205.

[66] Bracco P, Brunella V, Trossarelli L, Coda A, Botto-Micca F. Comparison of polypropylene and polyethylene terephthalate (Dacron) meshes for abdominal wall hernia repair: a chemical and morphological study. *Hernia* 2005; 9:51-55. 10.1007/s10029-004-0281-y.

[67] Ellison M. S, Fisher L. D, Alger K. W, Zeronian S. H. Physical properties of polyester fibers degraded by aminolysis and alkaline hydrolysis. *J Appl Polym Sci* 1982; 27:247-257. https://doi.org/10.1002/app.1982.070270126.

[68] Bu` I L. N, Thompson M, McKeown N. B, Romaschin A. D, Kalman P. G. Surface modification of the biomedical polymer poly (ethylene terephthalate). *Analyst* 1993; 118:463-474. https://doi.org/ 10.1039/AN9931800463.

[69] Fadeev A. Y, McCarthy J. Surface modification of poly (ethylene terephthalate) to prepare surfaces with silica-like reactivity. *Langmuir* 1998; 14:5586. https://doi.org/10.1021/la980512f.

[70] Avny Y, Rebenfeld L. Chemical modification of polyester fiber surfaces by amination reactions with multifunctional amines. *J Appl Polym Sci* 1986; 32:4009-4025. https://doi.org/10.1002/app.1986.070320318.

[71] Scharfstein J. S, Keaney Jr J. F, Slivka A, Welch G. N, Vita J. A, Stamler J. S, Loscalzo J. *J Clin Invest* 1994;94:1432-1439. https://doi.org/10.1172/JCI117480.

[72] Holowka E. P, Pochan D. J, Deming T. J. Charged polypeptide vesicles with controllable diameter. *J. Am. Chem. Soc.* 2005; 127:12423–12428. https://doi.org/10.1021/ja053557t.

[73] Kricheldorf H. R. Polypeptides and 100 years of chemistry of alpha-amino acid *N*- carboxyanhydrides. *Angew. Chem. Int. Ed. Engl.* 2006; 45:5752–5784. 10.1002/anie.200600693.

[74] Holmberg K, Bergström K, Brink C, Österberg E, Tiberg F, Harris J. M. Effects of protein adsorption, bacterial adhesion and contact angle of grafting PEG chains to polystyrene. *J. Adhes. Sci. Technol.* 1993; 7:503–517. https://doi.org/10.1163/156856193X00826.

[75] Lee J, Lee H, Andrade J. Blood compatibility of polyethylene oxide surfaces. *Prog. Polym. Sci.* 1995; 20:1043–1079. https://doi.org/10.1016/0079-6700(95)00011-4.

[76] Kogan G, S˘olte's L, Stern R, Gemeiner P. Hyaluronic acid: a natural biopolymer with a broad range of biomedical and industrial applications. *Biotechnol. Lett.*, 2007; 29: 17-25. 10.1007/s10529-006-9219-z.

[77] Lee J. Y, Spicer A. P. Hyaluronan: a multifunctional, megaDalton, stealth molecule. *Curr. Opin. Cell Biol.*, 2000; 12: 581-586. 10.1016/s0955-0674(00)00135-6.

[78] Xu X, Jha A. K, Harrington D. A, Farach-Carson M. C, Jia X. Hyaluronic Acid Based Hydrogels: From a Natural Polysaccharide to Complex Networks. *Soft Matter*, 2012; 8: 3280-3294. 10.1039/C2SM06463D.

[79] Lee H. E, Ahn H. Y, Mun J, Lee Y. Y., Kim M, Cho N. H, Chang K, Kim W. S, Rho J, Nam K. T. Amino-acid-and peptide-directed synthesis of chiral plasmonic gold nanoparticles, *Nature* 2018; 556: 360–365. https://doi.org/10.1038/s41586-018-0034-1.

[80] Song L, Wang S. F, Kotov N. A, Xia Y. S. Nonexclusive fluorescent sensing for L/D enantiomers enabled by dynamic nanoparticle-nanorod assemblies, *Anal. Chem.* 2012; 84: 7330–7335. https://doi.org/10.1021/ac300437v.

[81] Lin J, Huang B, Dai Y. F, Wei J. C, Chen Y. W. Chiral ZnO nanoparticles for detection of dopamine, *Mater. Sci. Eng.* C 2018; 93: 739–745. 10.1016/j.msec.2018.08.036.

[82] Ziyu Z, Feng Z, Jiaolong W, Xianhua Z, Xu W, Wu R, Lan L, Xiaolei W, Junchao W. L-cysteine modified ZnO: Small change while great progress. *Materials Science & Engineering* C 2019; 103:109818. https://doi.org/10.1016/j.msec.2019.109818.

[83] Renneberg R. High grade cysteine no longer has to be extracted from hair. In Demain, AL, eds. *Biotechnology for beginners*. Academic Press: Amsterdam 2008, pp. 106.

[84] Iorizzo M, Piraccini B. M, Tosti A. Nail cosmetics in nail disorders. *J Cosmet Dermatol* 2007; 6: 53-8. 10.1111/j.1473-2165.2007.00290.x.

[85] Hillebrand G, Bush R. D. Use of N-acatyl-L-cysteine and derivatives for regulating skin wrinkles and/or skin atrophy. *Great Britain Patent EP* 0 734 718 A2. 1992.

[86] Wacker Fermentation-Grade. Discovery a new dimension of pureness: vegetarian cysteine. Germany: Wacker Chemie AG. 2010.

[87] Zhou Y. T, He W, Lo Y. M, Hu X, Wu X, Yin J. J. Effect of silver nanomaterials on the activity of thiol-containing antioxidants. *J. Agric. Food Chem.* 2013; 61: 7855–7862. https://doi.org/10.1021/jf402146s.

[88] Tanabe S, Arai S, Watanabe M. Modification of wheat flour with bromelain and baking hypoallergenic bread with added ingredients. *Biosci. Biotech. Bioch.* 1996; 60: 1269– 1272. https://doi.org/10.1271/bbb.60.1269.

[89] Gomez-Juarez C, Casttelanos R, Ponce-Noyala T, Calderon V, Figueroa J. Protein recovery from slaughterhouse wastes. *Bioresource Technol*. 1999; 70: 129–133. https://doi.org/10.1016/S0960-8524(99)00030-9.

[90] Moure A, Dominguez H, Parajo, J. C. Fractionation and enzymatic hydrolysis of soluble protein present in waste from soy processing. *J. Agric. Food Chem.* 2005; 53: 7600–7608. https://doi.org/10.1021/jf0505325.

[91] Gupta V. K, Mohan D, Sharma S. Removal of lead from wastewater using bagasse fly ash—a sugar industry waste material. *Sep Sci Technol*, 1998; 33:1331–1343. https://doi.org/10.1080/01496399808544986.

[92] Song P, Guo X. Y, Pan Y. C, Shen S, Sun Y, Wen Y, Yang H.F. Insight in cysteamine adsorption behaviours on the copper surface by

electrochemistry and Raman spectroscopy. *Electrochim Acta*, 2013; 89:503–509. 10.1016/j.electacta.2012.11.096.

[93] Kuruvilla M, John S, Joseph A. Electrochemical studies on the interaction of L-cysteine with metallic copper in sulfuric acid. *Res Chem Intermed*, 2013; 39:3531–3543. 10.1007/s11164-012-0860-y.

[94] Eddy N. O. Experimental and theoretical studies on some amino acids and their potential activity as inhibitors for the corrosion of mild steel *J. Adv. Res.* 2011; 2: 35-47. https://doi.org/10. 1016/j.jare.2010. 08.005.

[95] Amin M. A, Khaled K. F, Mohsen Q, Arida H. A. A study of the inhibition of iron corrosion in HCl solutions by some amino acids. *Corrosion Science*, 2010; 52: 1684- 1695. https://doi.org/10.1016/j.corsci.2010.01.019.

[96] Saifi H, Bernard M.C, Joiret S, Rahmouni K, Takenouti H, Talhi B. Corrosion inhibitive action of cysteine on Cu–30Ni alloy in aerated 0.5 M H2SO4. *Materials Chemistry and Physics*, 2010; 120(2-3): 661-669. https://doi.org/10.1016/j.matchemphys.2009.12.011.

[97] Barouni K, Bazzi L, Salghi R, Mihit M, Hammouti B, Albourine A, El Issami S. Some amino acids as corrosion inhibitors for copper in nitric acid solution *Materials Letters*, 2008; 62(19): 3325-3327. 10.1016/j.matlet.2008.02.068.

[98] Kuruvilla M, Anupama R. P, Sam J, Abraham J. Enhanced Inhibition of the Corrosion of Metallic Copper Expose in Sulphuric Acid through the Synergistic Interaction of Cysteine and Alanine: Electrochemical and Computational Studies. *J Bio Tribo Corros* 2017; 3: 5. https://doi.org/10.1007/s40735-016-0064-x.

[99] Antonijevic M. M, Alagic S. C, Petrovic M. B, Radovanovic M. B, Stamenkovic A. T. The influence of pH on electrochemical behavior of copper in presence of chloride ions. *Int. J Electrochem Sci* 2009; 4:516–524.

[100] Ekemini B. I, Onyewuchi A, Saviour A. U. N-acetyl cysteine-based corrosion inhibitor formulations for steel protection in 15% HCl solution. *J. Mol. Liq.* 2017; 246: 112-118. https://doi.org/10. 1016/j.molliq.2017.09.040.

In: Cysteine: Sources, Uses and Health Effects ISBN: 978-1-53619-033-5
Editor: Taran Saunders © 2021 Nova Science Publishers, Inc.

Chapter 2

THERAPEUTIC USE OF N-ACETYLCISTEINE FOR OBSESSIVE COMPULSIVE DISORDER: A NEW AVENUE

Flavia di Michele[*], *MD, PhD*
Acute Psychiatric Unit, PTV Foundation-Policlinico Tor Vergata, Rome, Italy

ABSTRACT

Obsessive-compulsive disorder (OCD) is often a lifetime and disabling disorder since standard therapies both pharmacological and psychosocial may not be efficacious. A dysfunction of glutamatergic neurotransmission and neuroimmune abnormalities might be involved in the pathophysiology of this disorder. Lately, preclinical and clinical studies have shown a neuroprotective role for N-acetyl cysteine (NAC) by acting in two ways: as a modulator of inflammatory pathways and via the modulation of synaptic release of glutamate in cortico-subcortical brain regions.

---

[*] Corresponding Author's E-mail: flaviadimichele@gmail.com.

This chapter explores the therapeutic use of NAC in augmentation among individuals with refractory- OCD to the first- line pharmacological interventions, or in monotherapy, reviewing the published clinical studies.

The possible benefit mechanisms of NAC for this disorder will be discussed, given its specific property of stimulating the formation of glutathione in the brain.

Nutraceutical supplementation with NAC in OCD treatment may be important not only to improve psychiatric symptomatology, but also for its optimal safety and tolerability profile. This is of great interest especially considering the need for treatment special populations affected by OCD, such as children, youngsters and elders. Finally, the nutraceutical approach represents a good choice from a psychological perspective, given its better acceptance by the patients compared to pharmacological treatment.

Keywords: obsessive-compulsive disorder, N-Acetylcisteine, glutamate, neuroimmune

INTRODUCTION

Obsessive-compulsive disorder (OCD) is a complex mental illness with a lifetime prevalence of 1,6-3% in the general adult population. About 50% of all OCD cases appear during childhood or adolescence. The average age at onset in OCD patients is 19.5 years, with males manifesting a significantly younger age at onset than females [1, 2].

Its emergence in patients is associated with intrusive thought and bizarre ritualistic behavioral patterns, which can lead to significant impairments in personal, social, and occupational function for those afflicted. Therefore, depression, suicidality, functional impairment and days housebound are often associated to OCD. Patients with OCD impulsively and repeatedly think that they should do something regardless of their will. Typical obsessive behaviors include repeated check-ups of door locks, gas burning checks, hand washing, cleaning, walking without step on the line, symmetrically aligning objects, failing to discard objects, giving meaning to sounds, words and numbers, and sexual imagination [1, 3, 4].

Among individuals with OCD, the presence of another comorbid lifetime psychiatric disorder is around 90%. Anxiety disorders (75.8%), mood disorders (63.3%), and impulse control disorders (ICD) (55.9%) are considered the most frequent comorbidities [2]. Interestingly, OCD, ICD and substance-related disorders (SUD) have in common several features, namely: phenomenology, comorbidity, neuropathophysiology, neurocircuitry, and family history [5].

It has been shown that inappropriate and increased behavioral control due to an impairment of motor inhibition and altered cognitive flexibility, seem to depend upon the integrity of cortico-striato-thalamic circuit [6]. Accordingly, the progression and chronicity of OCD have been related to the involvement of more ventral striatal circuits [5].

Individuals with OCD might not benefit from standard therapies both pharmacological (serotonin selective reuptake inhibitors- SSRIs-, or clomipramine) and psychosocial interventions, such as 40% of subjects with OCD do not have a "true" clinical response or gain little benefit from these treatment strategies; therefore, those are called treatment-resistant [7].

Such a varied response to those treatment strategies may in part be explained by the heterogeneous neurobiology underlying the disorder, as well as to individual variances of pharmacodynamics and pharma-cokinetics mechanisms. Moreover, treatment adherence might be insufficient and this could be related to adverse effects associated to treatment with SSRIs, particularly at the required higher doses. This can be often a reason of treatment discontinuation [8].

The adjunctive use of neuroleptics to SSRIs is a current practice in treatment-resistant OCD [8].

Given the heterogeneity of OCD's neurobiology, theoretically, treatment of patients with this disorder should take into consideration the alterations in its underlying neurobiology.

Hence OCD patients in comorbidity with SUD or ICD might receive a combination treatment with agents deputed to the reduction or prevention of relapse of addiction (e.g., heavy drinking), which modulate the cortico-mesolimbic dopamine system through the opioid receptors (e.g., buprenorphine and naltrexone). Therapeutic agents modulating glutamate (e.g., topiramate, lamotrigine) may ameliorate compulsions, those acting on γ-aminobutyric acid (e.g., baclofen, clonazepam) system may be indicated for OCD associated with anxiety disorders, or tics. SSRIs should be considered when OCD is associated with depression [5].

The utility of glutamate-modulating drugs in add-on or in monotherapy in OCD has been recently reviewed. Memantine shows a positive effect as an augmentation therapy in OCD. Anticonvulsant drugs (lamotrigine, topiramate) and riluzole may also provide therapeutic benefit to some OCD patients. Ketamine exerts an important therapeutic potential due to a rapid onset of action [9].

However, besides the efforts, the current lack of reliable pharmacotherapies, still account for an enormous social burden linked to the chronic and disabling nature of this disorder [3] and strongly points out the need for further alternatives strategies.

Nutraceuticals are scarcely investigated in the psychiatric field, although represent promising alternatives, considering the critical need to identify safe and well tolerated treatments for OCD. Indeed, a huge proportion of OCD patients has to be treated in polytherapy, as already mentioned. This is not only because often monotherapy is unsuccessful, but also considering the treatment of comorbidities.

Another fundamental reason for promoting the use of nutraceutics as an alternative or supplementary therapy for OCD is the need for treating special populations, such as children and adolescents -aside to psychosocial interventions-. These OCD young patients have often to quit school or sport activities because of their difficulty in maintaining harmonious social relationships with peers and teachers.

Finally, addressing the treatment of the geriatric population, where the internistic comorbidities are very high and require special caution, represents another paramount issue for choosing nutraceutics. Indeed, elders are more susceptible to drug-induced adverse events.

This chapter explores the recent literature about clinical evidence for the therapeutic utility of N-acetylcysteine (NAC) that highlights this nutraceutical as a potential treatment option in OCD. It focuses on the property of NAC acting in a double manner as a glutamate-modulating agent and as modulator of inflammatory pathways.

The first part of the chapter addresses the NAC mechanisms of action and a general explanation of the pathogenetic hypothesis for OCD, highlighting the importance of the complex interplay between immune-inflammatory changes and neurocircuit dysfunctions resulting in alterations of the functional brain connectivity.

The second part of the chapter contains a review of the literature on the clinical data available concerning the efficacy of NAC treatment in OCD patients as an add-on therapy or in monotherapy.

The third part points out why choosing NAC for OCD treatment and conclusions are drawn.

NAC MECHANISMS OF ACTION

NAC, a derivate of the amino acid L-cysteine, is the antioxidant precursor of glutathione (GSH). The use of NAC in restoring GSH levels is well known [10]. GSH is the primary most important endogenous antioxidant in the brain, which neutralizes reactive oxygen and nitrogen species from the cell through both direct and indirect scavenging. Being ubiquitous, its role is to maintain the oxidative homeostasis in the cell [11].

After oral administration, NAC undergoes deacetylation in the liver to form cysteine, the substrate for GSH formation [12]. Unconverted cysteine enters the blood stream and it may cross the BBB via a sodium-dependent carrier and then converted into cystine [13]. Interestingly, in most regions of the brain, the uptake of glutamate and other anionic excitatory amino acids from the circulation is limited by the BBB. Therefore, the capacity of NAC to penetrate the BBB and to determinate the increase of brain GSH levels in animal models [14, 15] represents an important understanding in the psychiatric field, where alterations in brain GSH and other redox pathways have been demonstrated. Viceversa, oral administration of GSH does not adequately restore GSH brain levels, due to its rapid hydrolyzation by the liver and gut, with the consequence that the brain passage is poor [16].

Once NAC has been converted into cystine within the brain, high concentrations of cystine stimulate the exchange of intracellular glutamate for cystine through the cystine-glutamate antiporter, thereby elevating non-synaptic glutamate [17, 18]. This process activates the metabotropic glutamate receptors (mGluR2/3) on presynaptic neurons, inhibiting the synaptic release of glutamate and thereby restoring extracellular glutamate levels in the nucleus accumbens, which is part of the reward system [19, 20]. Thus, many studies have demonstrated that the modulation of this exchange system improves impulse control and reduces craving [21-23]. Intracellular cystine can then be reduced back to cysteine and used for GSH production, the potent and most abundant endogenous antioxidant in the body. Overall, NAC may regulate the exchange of glutamate, thus modulating the glutamatergic tone and preventing its pro-oxidant effects.

Beside to act as GSH precursor, NAC is able to scavenge oxidants directly, particularly the reduction of the hydroxyl radical,·OH and hypochlorous acid [24]. Therefore, the role of NAC in maintaining oxidative homeostasis is clear.

Another potential neuroprotective mechanism of NAC has been reported through its inhibitory effect on cerebral inflammatory response and on oxidative stress markers [25-27]. Moreover, NAC treatment protects the brain from inflammatory cytokines and this may be another potential mechanism by which NAC modulates the symptoms of psychiatric disorders [28]. Therefore, the crosstalk between antioxidant, anti-inflammatory and antiapoptotic mechanisms exerted by NAC sheds light on its potential neuroprotective role.

Due to all the mechanisms described, NAC has been increasingly studied for the treatment of OCD.

NAC PROPERTIES

NAC has been increasingly studied for the treatment of OCD disorders [for reviews see 29-31] as it may regulate the exchange of glutamate and prevent its pro-oxidant effects.

NAC properties are represented by greater bioavailability, stability, solubility and resistance to oxidation upon consumption than L-cysteine,

NAC exerts antioxidant, antiinflammatory, mucolytic and hepatoprotective properties [32, 33]. In the brain, NAC has demonstrated efficacy in counteracting oxidative stress, beside to normalization of glutamatergic transmission [33].

An advantage of NAC in comparison to other glutamatergic pharmaceuticals (Glutamate-modulating agents which have been investigated in OCD include riluzole, memantine, acamprosate, ketamine, d-cycloserine, and glycine – [9, 34]) is that it is available over-the-counter without prescription, and thus represents a more affordable and accessible treatment option. Further, being a nutraceutical, NAC meets a better acceptance by patients compared to pharmacological treatment.

THE PATHOGENETIC HYPOTHESIS FOR OCD

OCD, Glutamate and Oxidative Stress

Glutamatergic abnormalities have been largely demonstrated in OCD patients, but it still remains unclear if this is a primary, causal effect or a by-product of hypermetabolism and altered neurotransmission in other pathways [35]. Some studies have found abnormal glutamate metabolism in OCD subjects [36, 37]. Moreover, significantly higher levels of glutamate have been evidenced in the right caudate and orbitofrontal cortex, as well as in the cerebro-spinal fluid, of treatment refractory OCD patients [38-40].

On the other hand, low concentrations were found in the anterior cingulate in a sample of women with OCD, related with symptom severity [39].

It is well known that high levels of glutamate result in excitotoxicity and oxidative stress [41]. Oxidative stress is one of the possible pathological events that triggers the activation of degenerating cascades inside neuronal cells. Even more, the central nervous system is highly sensitive to oxidative stress due to low levels or capacities of scavengers [41].

Various studies have reported different types of oxidative stress in OCD populations, including increased lipid peroxidation [42-44], decreased vitamin E [45], catalase, glutathione peroxidase and selenium, increased superoxide dismutase [44] and changes in overall oxidative status [46]. Some of these alterations were correlated with the degree of symptomatology [43, 44].

Since the glutamatergic system mediates the acquisition and extinction of fear-conditioning, change in firing of the glutamatergic orbitofrontal neurons may play a role in the poor cognitive processing (impairment of cognitive flexibility) and doubt in OCD [47]. Besides, the role of glutamate in memory and cognition is well known [48, 49]. These data together with the pro-oxidant properties of glutamate support its suggested role in the pathophysiology of OCD. In favor of this hypothesis, pharmacological studies have shown the efficacy of glutamate modulating agents to regulate impulse control, a characteristic feature shared by obsessive-compulsive (OC) spectrum disorders, ICD and SUD [5, 9]. In this light, it is clear the new perspective of using glutamate-modulating agents in the treatment of OCD.

OCD and Autoimmune Diathesis

A growing body of evidence highlights the involvement of autoimmune mechanisms in the pathophysiology of a subset of OCD [50]. Alterations in pro- and anti-inflammatory cytokines, including interleukin (IL)-6 [51, 52], tumour necrosis factor (TNF)–α [52-54] and natural killer cells activity [53] have been documented in patients with OCD. These inflammatory cytokines are therefore implicated in the biological mechanisms underlying the pathophysiology of this disease. Moreover, the finding of increased CD8+ (e.g., suppressor T lymphocytes) decreased CD4+ (helper t lymphocytes) in adult OCD subjects compared with healthy controls has been demonstrated [55].

Adult OCD is often in comorbidity with systemic diseases, namely: thyroid autoimmune disease, rheumatoid arthritis, lupus eritematosus systemic, multiple sclerosis, etc. [56-60]. In pediatric OCD humoral immunity seems to be strongly involved in its pathophysiology. Primary immune responses to streptococcal infections may play an important role in the etiology of pediatric OCD [61, 62]. The evidence of a good response to immunomodulatory therapies in children with treatment refractory OCD [63] supports the neuroimmune hypothesis for some sub-sets of OCD patients. Moreover, it has been reported that the proportion of OCD patients with positive Anti-streptolysin -ASO- titre (>200 IU/ml) was significantly higher than in major depression patients. Furthermore, in a small proportion of them anti-brain antibodies were documented, indicating an autoimmune reaction [64]. ASO titration may be considered as a marker in the context of autoimmune pathologies, including OCD occurring after streptococcal infections [65]. Immune deficits in OCD are discussed with recommendations for surveillance of patient's immune status. Simple laboratory analyses (leucocytic formula, C-reactive protein, Rheuma test, ASO, etc.) suggestive of the presence of an inflammatory process, together with a complete thyroid function tests, have been indicated in OCD patients, in search for possible associations with autoimmune systemic diseases, due to possible immunologic cross-reactivity [66]. This could identify a subgroup of OCD patients, confirming the hypothesis of a heterogeneity in the neurobiology of OCD.

NAC IN OCD TREATMENT

Given the mentioned properties and mechanisms of action of NAC such as the modulation of the glutamatergic tone and the reduction of oxidative stress [33], as well as its anti-inflammatory action, this compound has been increasingly studied for OCD treatment.

A first preclinical design showed a dose dependent reduction of marble burying behavior after NAC administration [67]. Another preclinical study demonstrated that NAC at different dosages (60 and 120 mg/kg/day) was capable to block 5-HT1B agonist-induced OCD-related behavior in mice. In a separate study, the administration of NAC at 60 mg/kg/day was able to block 5-HT1B agonist-induced OCD-like behavior after 3 weeks, but not after 1 week of NAC treatment [68].

In the last two decades, case reports and clinical trials of NAC in the treatment OCD and OC related disorders have been published. From the reviewed literature, data support the use of NAC supplementation in cases of moderate- severity OCD. Doses of 2000-3000 mg/day for a minimum of eight weeks (preferably 12 weeks) may be adequate for exerting an efficacious therapeutic effect [69-71; 73-80].

The first case-report on the use of NAC in augmentation with SSRIs evidenced an important benefit in a treatment resistant OCD patient. The subject who gained a partial benefit from treatment with fluvoxamine (300 mg/day), continued fluvoxamine in add-on with 3,000 mg of NAC for a period of 13 weeks trial (including dose titration to 3,000 mg). During this period, the patient improved on both the assessment scales: Yale–Brown Obsessive Compulsive Scale (Y-BOCS) and Hamilton Rating Scale for Depression (HAM-D) scores. At the end of the trial significant improvements in control of compulsive washing and obsessional triggers were evidenced, and the Y-BOCS score was reduced to 9 [69].

A second case report on the use of NAC in add-on with paroxetine (60 mg/day) and risperidone (2 mg/day) showed a complete remission of all symptoms in a 44-year-old patient with resistant OCD, using NAC at a dosage of 1,200 mg/die. The duration of the study was not specified. No adverse side effects were reported [70].

The first randomized clinical trial evaluating the use of NAC in add-on treatment of OCD [71] was a 12-weeks trial, where NAC was titrated from 600 mg/day and doubled weekly up to dose of 2,400 mg/day (at week-3), then maintained for the duration of the trial. Y-BOCS and Clinical Global Improvement scales (CGI-S) at four-week intervals were used for assessment. The treatment group (n = 20) demonstrated a significant improvement compared to placebo (n = 19) according to the Y-BOCS scores. Specifically, a gradual but continuous decrease in symptom severity appeared from week-4, with the NAC group showing significance over placebo from week-8 onwards. However, only half of the treatment group recorded a full clinical response. This is in line with the hypothesis of a heterogeneous etiology of the disorder and that NAC may be effective only in a subset of OCD patients. Unfortunately, comorbidities were not considered in the trial, these could have identified whether NAC was more beneficial in patients with less complex presentations. The severity of this particular group of OCD participants was moderate (mean Y-BOCS = 27), representing a select population of OCD patients [71]. Finally, a longer clinical trial may have determined if NAC efficacy would have continued, as shown in clinical trials involving bipolar depressive patients who obtained continual improvements in their depressive symptoms occurring over a 24 weeks period [72].

Vice versa, a series of six retrospective case reports concluded for a poor response to an add-on NAC treatment of OCD [73]. All six patients presented with severe OCD (Y-BOCS score of 29.3 ± 4.3) were treatment-resistant (i.e., unresponsive to at least two first line pharmacotherapies). The NAC dose was titrated to 3,000 mg/day, but titration lengths were not reported. Most of them were treated with 3,000 mg/day for a month. NAC was used adjunctively in this series and all patients were receiving pharmacotherapies at doses which had been stable for a minimum of eight weeks prior to supplementation with NAC. The only patient who demonstrated a modest response to NAC received a diagnosis of 'moderate' OCD severity (Y-BOCS of 26 baseline, 17 at endpoint), and major depressive disorder was the only identified comorbidity, as compared to other patients who experienced multiple comorbidities. Interestingly, the two patients who were non-responsive to glutamatergic medications (topiramate and lamotrigine) demonstrated a worsening in their symptom severity according to their Y-BOCS score after NAC supplementation. This is in line with the hypothesis of the heterogeneous nature of OCD neurobiology and may indicate that NAC could not be efficacious in glutamate-independent presentations of OCD.

A 10 week- placebo-controlled trial, assessed the efficacy and tolerability of an add-on NAC treatment in moderate-to-severe OCD. In this study, 44 adult patients were randomized into two parallel groups to receive fluvoxamine (200 mg daily) plus placebo or fluvoxamine (200 mg daily) plus NAC (2000 mg daily) up to 10 weeks. Significant effects for time × treatment interaction in the Y-BOCS total score and a significant effect for time × treatment interaction in the Y-BOCS obsession subscale between the two groups were evidenced. Hence, the authors claimed that NAC in add-on therapy may be of benefit in mild to severe OCD [74].

A randomized double-blind, placebo-controlled, 16-weeks trial of NAC (3,000 mg daily) in adults (aged 18-65 years) with treatment-resistant OCD [75], as well as a previous analogue designed study in 44 adult participants with OCD [76], did not demonstrate a significant benefit efficacy of NAC in reducing OCD severity in treatment-resistant OCD adults. However, in the latest study [75], a secondary analysis evidenced that NAC might be useful in reducing anxiety symptoms in these treatment-resistant OCD patients. Even in the previous 16 weeks trial [76], although the primary outcome measure Y-BOCS revealed a not significant time × treatment interaction for the Y-BOCS scale total score, a significant time × treatment interaction was observed for the Y-BOCS 'Compulsions' subscale in favor of NAC ($p = 0.013$), with a significant reduction observed at week 12 (dissipating at week 16).

The first case report on an adolescent treatment-resistant OCD showed the recovery achieved by adding NAC to citalopram (60 mg/day). NAC was titrated from 600 mg to 2400 mg daily over six weeks and maintained for further 18 weeks, while treatment with 60 mg/day citalopram was continued. In 24 weeks, a significant decrement in symptom severity was reported by the Y-BOCS, with a baseline score of 37 reduced to 9 [77].

The same authors have published a series of 5 case reports, also in pediatric age. All patients were initially treated with NAC 600 mg daily, in add-on therapy. The NAC dosage was then titrated up, to a final dosage that varied from 1,800 mg to 3,000 mg. Four of the 5 patients had significant improvement with NAC augmentation therapy, after 24 weeks follow-up. Two of them experienced a full response to treatment, the other two patients entered remission, considered when Y-BOCS score is less than 16. The remaining patient, considered as a "non-responder", had a decrease in Y-BOCS score less than 25% and CGI-S of 4 [78].

A 10 weeks randomized double-blind, placebo-controlled clinical trial, including 34 children and adolescents (from 10 to 21 years old) OCD outpatients was carried out. Their patients received SSRIs, mainly citalopram (20 to 40 mg/day) plus NAC (titred up to 2,400 mg/day) (16 patients) or placebo (15 patients). Three other patients received fluoxetine (20 mg/day) or fluvoxamine (100 mg/day) or sertraline (50 mg/day) plus NAC. The Y-BOCS score of NAC group significantly decreased, while no statistically significant decrease of Y-BOCS was found in the placebo group. Particularly the score of resistance/control to compulsion was improved in the NAC group. The score of three domains of quality of life significantly decreased in NAC group compared to placebo group. No serious adverse effect was found in the two groups. Thus, this study demonstrates that NAC increases the effect of citalopram by improving resistance/control to compulsions in OCD children and adolescents. In addition, it shows a good tolerability profile [79].

Interestingly, a very recent pilot study on children and adolescents (aged 8-17) demonstrated that even NAC alone, at a dosage up to 2,700 mg/day, is able to reduce the severity of OC symptoms, and shows a good tolerability profile [80].

NAC Tolerability Profile

NAC is considered a compound very well tolerated, with minimal side effects reported [31]. The most common reported are mild gastrointestinal effects including flatulence, a transitory side effect which disappears within two weeks from the start of NAC treatment. Following by symptoms of headache, skin rash, diarrhea and nausea and vomiting of mild-moderate intensity [32]. However, nausea and vomiting were also reported in the placebo group of one particular trial, and was not evident in others [71]. Another study has reported no adverse effects and a good tolerance of NAC in doses up to 8,000 mg/day [73]. This should encourage psychiatrists to use higher dose of NAC in OCD disorders who appear resistant to doses of 3,000 mg/day used in the discussed clinical trials.

WHY CHOOSING NAC FOR OCD TREATMENT

OCD is a leading cause of disability, being associated with considerable burden of disease, functional impairment, social burden, family burden, physical comorbidities, suicidality, high economic costs, and poor quality of life [1-3]. Epidemiological studies showed that across anxiety disorders OCD represents the one with the highest estimate of the number of life years lost due to the disease in men and second highest in women behind panic disorder [81].

One of the unmet needs for OCD is represented by the inadequate response to pharmacological treatment and to other psychosocial interventions. A second unmet need for OCD is considered the safety and the tolerability of the antidepressant therapy. Patients frequently discontinue treatment with SSRIs due to common adverse events (sexual side effects, sedation, weight gain, nausea, etc.) which impair their quality of life [8]. Moreover, a polytherapy is often necessary for OCD itself and/or for comorbidities associated to it, either internistic or psychiatric ones. Therefore, interactions among drugs are a fundamental issue to take into account in a treatment's choice. Antidepressants act on liver cythochromes, caution is hence required when associated with drugs that are metabolized by the same cytochromes.

Comorbidity with other psychiatric conditions is up to 90%, while comorbidity with chronic medical conditions in OCD is especially high in the elderly [2, 81]. Elders are more susceptible to drug-induced adverse events. Among the most troublesome ones: the anticholinergic effects, namely, constipation, sedation, urinary retention, delirium and cognitive dysfunctions. Even the anti-adrenergic effects of antidepressants are difficult to manage in elders: in particular the postural hypotension, orthostatic hypotension, and sedation can lead to falls and bone fractures. Even more, considering that Antidepressants reduce the bone mineral density [82], treating OCD in the elderly represents a very critical issue for all the mentioned reasons.

Analogously, the necessity for a pharmacological treatment in the young population of OCD represents another unmet need.

Therefore, the use of NAC appears to be very useful, considering the limitations of the use of traditional pharmacological treatment for these populations of patients.

The association between OCD, inflammation and immune activation evidenced in the last years, has determined to target inflammatory pathways as a therapeutic tool for individuals with OCD [31]. In particular, a subset of OCD patients, demonstrates a dysregulated immune function with a unique pattern of increased inflammatory mediators, chemokines and colony-stimulating factors, providing support for the immune hypothesis of OCD [51-57].

Moreover, the dysregulation of brain neurotransmitters, with alterations in signaling and metabolism of glutamate in patients with OCD suggests novel treatment targets, as an alternative to the antidepressants acting on the monoamine system [9]. The emergence of the NMDA receptor antagonist, ketamine, as a rapidly acting antidepressant is one of the most lately studied [34]. Unfortunately, ketamine shows several drawbacks such as psychotomimetic, dissociative symptoms, abuse potential and neurotoxicity, all of which represent a huge limitation for its use, that require much caution and prevent its routine use in daily clinical practice [83, 84].

In parallel, NAC is a natural compound, a redox-active GSH precursor that could modulate glutamatergic neurotransmission, restoring the imbalance between excitatory and inhibitory elements of the fronto-subcortical neurocircuitry. The modulation of inflammatory pathways decreasing inflammatory cytokines may also play a role in the benefits seen following NAC treatment. As a modulator of the systems described, NAC promotes neuroprotection, neurogenesis, and decreases apoptosis, all of which contribute to the neurobiology of OCD. Besides, NAC is an antioxidant agent with minimal side effects.

Given the links between glutamate, inflammation and OCD, the interest in NAC who targets this cross-talk, but differently from drugs, such as ketamine, is safer and well tolerated (having no side effects or minimal side effects) is clear.

Therefore, NAC may be a therapeutic tool for treating psychiatric disorders involving glutamatergic dysfunction and inflammatory conditions, such as OCD.

The large body of clinical trials in the treatment of OCD and OC spectrum disorders, and more ongoing research, testifies NAC to be a relatively new potential complementary treatment of these disorders. Indeed, NAC has been shown to improve OCD symptomatology with a good tolerability profile and minimal adverse effects. However, the current evidence for the real efficacy of NAC as treatment for OCD still remains inconsistent. The heterogeneity in the results of these studies derives from methodological problems, i.e., the use of small sample sizes, short durations, differing baseline symptomatology and poorly defined illness durations, not defined comorbidities, which makes it impossible to provide recommendations on the routine clinical use of NAC in the treatment of OCD. This addresses the need for further rigorous clinical investigations.

Moreover, given the heterogeneous nature of OCD neurobiology [5], NAC may be effective toward OCD subgroups involving glutamatergic dysfunction, or associated with an autoimmune diathesis, but ineffective against glutamate-independent presentations of OCD. Therefore, it may help to identify different subtypes of the pathology.

Finally, comorbidity in OCD and a lifetime history of suicide attempt are common [1] and represent an overload of disability. Thus, an alternative approach to OCD treatment may be important not only for the improvement of the symptomatology, but also for reducing the psychological burden linked to the stigma associated with this psychiatric condition. Indeed, the supplementation with nutraceuticals such as NAC addresses the complex issue of OCD from a new point of view, where restoring levels of natural anti-inflammatory molecules may represent the key element for the improvement of the clinical features. Finally, nutraceutical treatment seems to meet a better acceptance from the patients, compared to pharmacological treatment [85]. This is not a trivial issue since therapeutic adherence can be one of the reasons for inadequate response to drugs.

Last but not least, the reduction of OC symptoms would secondly improve the cognition and social functioning of these patients, ameliorating their quality of life and reducing the costs of OCD.

CONCLUSION

Considering the inadequate response to pharmacological treatment and to other psychosocial interventions, aside the frequent requirement of a polytherapy for OCD treatment itself, and for comorbidities associated to it, either internistic or psychiatric ones, interaction among different drugs has to be addressed, particularly considering special populations, such as elders as well as young patients in OCD.

The robust association between OCD, inflammation and immune activation evidenced in the last years, has determined to target inflammatory pathways as a therapeutic tool for individuals with OCD. NAC exerts antioxidant and anti-inflammatory actions by reduction of inflammatory cytokines and oxidative stress. It also has a neuroprotective role against excitotoxicity due to glutamate alterations, where contribution of inflammatory cytokines has been demonstrated.

For all of these reasons, nutraceutical supplementation with NAC in OCD treatment may represent an important alternative to the polytherapy with SSRIs plus neuroleptics, being free from adverse side effects and improving OC symptomatology.

The purpose of this review was thus to illustrate the role of NAC as a novel strategy for OCD treatment, to describe the potential mechanisms of action, the available clinical evidence of its efficacy in this field, and to better inform clinicians about the possible benefits of using NAC supplementation in OCD. Finally, considering the problem of therapeutic adherence in these patients, the nutraceutical approach represents a good choice from a psychological perspective, given its better acceptance by the patients compared to pharmacological treatment.

REFERENCES

[1] Brakoulias V, Starcevic V, Belloch A, et al. Comorbidity, age of onset and suicidality in obsessive-compulsive disorder (OCD): An international collaboration. *Compr Psychiatry* 2017; 76: 79-86.

[2] Ruscio AM, Stein DJ, Chiu WT, Kessler RC. The epidemiology of obsessive-compulsive disorder in the National Comorbidity Survey Replication. *Mol Psychiatry* 2010; 15: 53–63.

[3] DuPont RL, Rice DP, Shiraki S, Rowland CR. Economic costs of obsessive-compulsive disorder. *Medical interface* 1995; 8: 102-109.

[4] First, MB. *Diagnostic and Statistical Manual of Mental Disorders*. 5th ed. Washington, DC: American Psychiatric Association; 2013.

[5] Fontenelle LF, Oostermeijer S, Harrison BJ, Pantelis C, Yücel M. Obsessive-compulsive disorder, impulse control disorders and drug addiction: common features and potential treatments. *Drugs* 2011; 71:827-840.

[6] Chamberlain SR, Fineberg NA, Blackwell AD, Robbins TW, Sahakian BJ. Motor inhibition and cognitive flexibility in obsessive-compulsive disorder and trichotillomania. *Am J Psychiatry* 2006; 163: 1282-1284.

[7] Pallanti S, Grassi G, Cantisani A. Emerging drugs to treat obsessive-compulsive disorder. *Expert Opin Emerg Drugs* 2014; 19: 67-77.

[8] Hirschtritt ME, Bloch MH, Mathews CA. Obsessive-Compulsive Disorder: Advances in Diagnosis and Treatment. *JAMA* 2017; 317(13): 1358-1367.

[9] Marinova Z, Chuang DM, Fineberg N. Glutamate-modulating drugs as a potential therapeutic strategy in obsessive-compulsive disorder. *Curr Neuropharmacol* 2017; 15(7): 977-995. doi: 10.2174/1570159 X1566617032010423.

[10] Scalley RD, Conner CS. Acetaminophen poisoning: a case report of the use of acetylcysteine. *Am J Hosp Pharm* 1978; 35: 964-7.

[11] Dringen R, Hirrlinger J. Glutathione pathways in the brain. *Biol Chem* 2003; 384: 505-16.

[12] Atkuri KR, Mantovani JJ, Herzenberg LA, Herzenberg LA. N-Acetylcysteine--a safe antidote for cysteine/glutathione deficiency. *Curr Opin Pharmacol* 2007; 7: 355-359.
[13] Smith QR. Transport of glutamate and other amino acids at the blood-brain barrier. *J Nutr* 2000; 130(4S Suppl): 1016S- 1022S.
[14] Dean O, van den Buuse M, Copolov D, et al. N-acetylcysteine inhibits depletion of brain glutathione levels in rats: implications for schizophrenia. *Int J Neuropsychopharmacol* 2004; 7(S1): 262.
[15] Farr SA, Poon HF, Dogrukol-Ak D, et al. The antioxidants alphalipoic acid and N-acetylcysteine reverse memory impairment and brain oxidative stress in aged SAMP8 mice. *J Neurochem* 2003; 84: 1173-83.
[16] Witschi A, Reddy S, Stofer B, et al. The systemic availability of oral glutathione. *Eur J Clin Pharmacol* 1992; 43: 667-9.
[17] Kau KS, Madayag A, Mantsch JR, Grier MD, Abdulhameed O, Baker DA. Blunted cystine-glutamate antiporter function in the nucleus accumbens promotes cocaine-induced drug seeking. *Neuroscience* 2008; 155: 530-537.
[18] Kalivas PW. The glutamate homeostasis hypothesis of addiction. *Nat Rev Neurosci* 2009; 10: 561-572.
[19] Baker DA, Shen H, Kalivas PW. Cystine/glutamate exchange serves as the source for extracellular glutamate: modifications by repeated cocaine administration. *Amino Acids* 2002; 23: 161-162.
[20] Moran MM, McFarland K, Melendez RI, Kalivas PW, Seamans JK. Cystine/glutamate exchange regulates metabotropic glutamate receptor presynaptic inhibition of excitatory transmission and vulnerability to cocaine seeking. *J Neurosci* 2005; 25: 6389-6393.
[21] Baker DA, McFarland K, Lake RW, Shen H, Toda S, Kalivas PW. N-acetyl cysteine-induced blockade of cocaine induced reinstatement. *Ann N Y Acad Sci* 2003; 1003: 349-351.
[22] Kupchik YM, Moussawi K, Tang XC, Wang X, Kalivas BC, Kolokithas R, et al. The effect of N-acetylcysteine in the nucleus accumbens on neurotransmission and relapse to cocaine. *Biol Psychiatry* 2012; 71: 978-986.

[23] Schmaal L, Veltman DJ, Nederveen A, van den Brink W, Goudriaan AE. N-acetylcysteine normalizes glutamate levels in cocaine-dependent patients: a randomized crossover magnetic resonance spectroscopy study. *Neuropsychopharmacology* 2012; 37: 2143-2152.

[24] Bauer J, Pedersen A, Scherbaum N, et al. Craving in alcohol-dependent patients after detoxification is related to glutamatergic dysfunction in the nucleus accumbens and the anterior cingulate cortex. *Neuropsychopharmacology* 2013; 38: 1401-1408.

[25] Aruoma OI, Halliwell B, Hoey BM, et al. The antioxidant action of N-acetylcysteine: its reaction with hydrogen peroxide, hydroxyl radical, superoxide, and hypochlorous acid. *Free Radic Biol Med* 1989; 6: 593-7.

[26] Khan M, Sekhon B, Jatana M, et al. Administration of N-acetyl-cysteine after focal cerebral ischemia protects brain and reduces inflammation in a rat model of experimental stroke. *J.Neurosci Res* 2004; 76: 519-27.

[27] Chen G, Shi J, Hu Z et al. Inhibitory effect on cerebral inflammatory response following traumatic brain injury in rats: a potential neuroprotective mechanism of N-acetylcysteine. *Mediators Inflamm* 2008; 8: 716458.

[28] Nascimento MM, Suliman ME, Silva M et al. Effect of oral N-acetylcysteine treatment on plasma inflammatory and oxidative stress markers in peritoneal dialysis patients: a placebo-controlled study. *Perit Dial Int* 2010; 30: 336-42.

[29] Oliver G, Dean O, Camfield D, Blair-West S, Berk M, Sarris J. N-acetyl cysteine in the treatment of obsessive compulsive and related disorders: a systematic review. *Clin Psychopharmacol Neurosci* 2015; 13: 12-24.

[30] Couto JP, Moreira R. Oral N-acetylcysteine in the treatment of obsessive-compulsive disorder: A systematic review of the clinical evidence. *Prog Neuropsychopharmacol Biol Psychiatry* 2018; 86: 245-254.

[31] di Michele F, Siracusano A, Talamo A, Niolu C. N-Acetyl Cysteine and Vitamin D Supplementation in Treatment Resistant Obsessive-

compulsive Disorder Patients: A General Review. *Curr Pharm Des.* 2018;24(17):1832-1838.

[32] Bonanomi L, Gazzaniga A. Toxicological, pharmacokinetic and metabolic studies on acetylcysteine. *Eur J Respir Dis Suppl.* 1980; 111: 45-51.

[33] Berk M, Malhi GS, Gray LJ, Dean OM. The promise of N-acetylcysteine in neuropsychiatry. *Trends Pharmacol Sci* 2013; 34: 167-177.

[34] Ballard ED, Ionescu DF, Vande Voort JL, Niciu MJ, Richards EM, Luckenbaugh DA, Brutsch NE, Ameli R, Furey ML, Zarate CA, Jr. Improvement in suicidal ideation after ketamine infusion: relationship to reductions in depression and anxiety. *J Psychiatr Res* 2014; 58: 161–166.

[35] Carlsson ML. On the role of prefrontal cortex glutamate for the antithetical phenomenology of obsessive compulsive disorder and attention deficit hyperactivity disorder. *Prog Neuropsychopharmacol Biol Psychiatry* 2001; 25: 5-26.

[36] Pittenger C, Bloch MH, Williams K. Glutamate abnormalities in obsessive compulsive disorder: neurobiology, pathophysiology, and treatment. *Pharmacol Ther* 2011; 132: 314-332.

[37] Wu K, Hanna GL, Rosenberg DR, Arnold PD. The role of glutamate signaling in the pathogenesis and treatment of obsessive-compulsive disorder. *Pharmacol. Biochem Behav* 2012; 100:726-735.

[38] Bhattacharyya S, Khanna S, Chakrabarty K, Mahadevan A, Christopher R, Shankar SK. Anti-brain autoantibodies and altered excitatory neurotransmitters in obsessive-compulsive disorder. *Neuropsychopharmacology* 2009; 34: 2489-2496.

[39] Yücel M, Wood SJ, Wellard RM, et al. Anterior cingulate glutamate-glutamine levels predict symptom severity in women with obsessive-compulsive disorder. *Aust N Z J Psychiatry* 2008; 42: 467-477.

[40] Chakrabarty K, Bhattacharyya S, Christopher R, Khanna S. Glutamatergic dysfunction in OCD. *Neuropsychopharmacology* 2005; 30: 1735-1740.

[41] Burdo J, Dargusch R, Schubert D. Distribution of the cystine/glutamate antiporter system xc- in the brain, kidney, and duodenum. *J Histochem Cytochem* 2006; 54: 549-557.

[42] Kuloglu M, Atmaca M, Tezcan E, et al. Antioxidant enzyme activities and malondialdehyde levels in patients with obsessive-compulsive disorder. *Neuropsychobiology* 2002; 46: 27-32.

[43] Chakraborty S, Singh OP, Dasgupta A, et al. Correlation between lipid peroxidation-induced TBARS level and disease severity in obsessive-compulsive disorder. *Prog Neuropsychopharmacol Biol Psychiatry* 2009; 33: 363-6.

[44] Ozdemir E, Cetinkaya S, Ersan S, Kucukosman S, Ersan EE. Serum selenium and plasma malondialdehyde levels and antioxidant enzyme activities in patients with obsessive-compulsive disorder. *Prog Neuropsychopharmacol Biol Psychiatry* 2009; 33: 62-5.

[45] Ersan S, Bakir S, Erdal Ersan E, Dogan O. Examination of free radical metabolism and antioxidant defence system elements in patients with obsessive-compulsive disorder. *Prog Neuropsycho-pharmacol Biol Psychiatry* 2006; 30: 1039-42.

[46] Selek S, Herken H, Bulut M, et al. Oxidative imbalance in obsessive compulsive disorder patients: a total evaluation of oxidant-antioxidant status. *Prog Neuropsychopharmacol Biol Psychiatry* 2008; 32: 487-91.

[47] Riaza Bermudo-Soriano C, Perez-Rodriguez MM, Vaquero- Lorenzo C, Baca-Garcia E. New perspectives in glutamate and anxiety. *Pharmacol Biochem Behav* 2012; 100: 752-774.

[48] Morris RG. NMDA receptors and memory encoding. *Neuropharmacology* 2013; 74: 32-40.

[49] Henley JM1, Wilkinson KA. AMPA receptor trafficking and the mechanisms underlying synaptic plasticity and cognitive aging. *Dialogues Clin Neurosci* 2013; 15: 11-27.

[50] di Michele F. Utility of systematic studies of the immune function in obsessive-compulsive disorder patients. *Aust N Z J Psychiatry* 2007; 41: 460-461.

[51] Carpenter LL, Heninger GR, McDougle CJ, et al. Cerebrospinal fluid interleukin-6 in obsessive compulsive disorder and trichotillomania. *Psychiatry Res* 2002; 112: 257-262.

[52] Fluitman S, Denys D, Vulink N, et al. Lipopolysaccharide induced cytokine production in obsessive-compulsive disorder and generalized social anxiety disorder. *Psychiatry Res* 2010; 178: 313–316.

[53] Denys D, Fluitman S, Kavelaars A, et al. Decreased TNF-alpha and NK activity in obsessive compulsive disorder. *Psychoneuroendocrinology* 2004; 29: 945-952.

[54] Monteleone P, Catapano F, Fabrazzo M, et al. Decreased blood levels of tumor necrosis factor-alpha in patients with obsessive compulsive disorder. *Neuropsychobiology* 1998; 37: 182-185.

[55] Marazziti D, Presta S, Pfanner C, et al. Immunological alterations in adults obsessive-compulsive disorder. *Biol Psychiatry* 1999; 46: 810-814.

[56] Hantouche E, Piketty ML, Poirier MF, et al. Obsessive-compulsive disorder and the study of thyroid function. *Encephale* 1991; 17: 493-6.

[57] Dinn WM, Harris CL, McGonigal KM, Raynard RC. Obsessive-Compulsive disorder and immunocompetence. *Int J Psychiatry Med* 2001; 31: 311-320.

[58] Slattery MJ, Dubbert BK, Allen AJ, et al. Prevalence of Obsessive-compulsive disorder in patients with systemic lupus erythematosus. *J Clin Psychiatry* 2004; 65: 301-306.

[59] Miguel EC, Stein MC, Rauch S, et al. Obsessive-compulsive disorder in patients with multiple sclerosis. *J Neuropsychiatry Clin Neurosci* 1995; 7: 507-510.

[60] Witthauer C, Gloster AT, Meyer AH, Lieb R. Physical diseases among persons with obsessive compulsive symptoms and disorder: a general population study. *Soc Psychiatry Epidemiol* 2014; 49: 2013-2022.

[61] Swedo SE, Leonard HL, Garvey M, et al. Pediatric autoimmune neuropsychiatric disorders associated with streptococcal infections:

clinical description of the first 50 cases. *Am J Psychiatry* 1998; 155: 264-71.

[62] Arnold PD, Richter MA. Is obsessive-compulsive disorder an autoimmune disease? *CMAJ* 2001; 165: 1353-8.

[63] Perlmutter SJ, Leitman SF, Garvey MA, et al. Therapeutic plasma exchange and intravenous immunoglobulin for obsessive-compulsive disorder and tic disorders in childhood. *Lancet* 1999; 354: 1153-1158.

[64] Maina G, Albert U, Bogetto F, et al. Anti-brain antibodies in adult patients with obsessive-compulsive disorder. *J Affect Dis* 2009; 116: 192-200.

[65] Pozzi M, Pellegrino P, Carnovale C, et al. On the connection between autoimmunity, tic disorders and obsessive-compulsive disorders: a meta-analysis on anti-streptolysin O titres. *J Neuroimmune Pharmacol* 2014; 9: 606-14.

[66] di Michele F. Improving the management of Obsessive-compulsive disorder by considering the autoimmune diathesis. *Aust N Z J Psychiatry* 2017; 51(11): 1159-1160. doi: 10.1177/0004867417 715915.

[67] Egashira N, Shirakawa A, Abe M, et al. N-acetyl-L-cysteine inhibits marble- burying behavior in mice. *J Pharmacol Sci* 2012; 119: 97-101.

[68] Allen EM, Hughes EF, Anderson CJ, Woehrle NS. N-acetylcysteine blocks serotonin 1B agonist-induced OCD-related behavior in mice. *Behav Neurosci* 2018; 132(4): 258-268. doi: 10.1037/bne0000251.

[69] Lafleur DL, Pittenger C, Kelmendi B, et al. N-acetylcysteine augmentation in serotonin reuptake inhibitor refractory obsessive compulsive disorder. *Psychopharmacology* 2006; 184: 254-6.

[70] Saraiva, S., Jesus, G., Gonçalves, F. & Mota, T. Complete Remission of Obsessive Compulsive Disorder After N-acetylcystein Treatment. *Eur Psychiatry* 2015; 3: 1499.

[71] Afshar, H, Roohafza H, Mohammad-Beigi H, Haghighi M, Jahangard L, Shokouh P, et al. N-acetylcysteine add-on treatment in refractory obsessive-compulsive disorder: a randomized, double-blind, placebo-controlled trial. *J Clin Psychopharmacol* 2012; 32: 797-803.

[72] Berk M, Copolov DL, Dean O, et al. N-acetyl cysteine for depressive symptoms in bipolar disorder--a double-blind randomized placebo-controlled trial. *Biol Psychiatry* 2008; 64: 468-475.

[73] Van Ameringen M, Patterson B, Simpson W, Turna J. N-acetylcysteine augmentation in treatment resistant obsessive compulsive disorder: A case series. *J Obsessive-Compuls Relat Disord* 2013; 2: 48-52.

[74] Paydary K, Akamaloo A, Ahmadipour A, et al. N-acetylcysteine augmentation therapy for moderate-to-severe obsessive-compulsive disorder: randomized, double-blind, placebo-controlled trial. *J Clin Pharm Ther* 2016; 41: 214-9.

[75] Costa DLC, Diniz JB, Requena G, et al. Randomized, Double-Blind, Placebo-Controlled Trial of N-Acetylcysteine Augmentation for Treatment-Resistant Obsessive-Compulsive Disorder. *J Clin Psychiatry* 2017; 78(7): e766-e773. doi: 10.4088/JCP.16m11101.

[76] Sarris J, Oliver G, Camfield DA, et al. N-Acetyl Cysteine (NAC) in the Treatment of Obsessive-Compulsive Disorder: A 16-Week, Double-Blind, Randomized, Placebo-Controlled Study. *CNS Drugs* 2015; 29: 801-9.

[77] Yazici, KU, Percinel I. The role of glutamatergic dysfunction in treatment resistant obsessive-compulsive disorder: treatment of an adolescent case with N-acetylcysteine augmentation. *J child adolesc psychopharmacol* 2014; 24: 525–527.

[78] Yazici, KU, Percinel I. N-Acetylcysteine Augmentation in Children and Adolescents Diagnosed With Treatment-Resistant Obsessive-Compulsive Disorder: Case Series. *J Clin Psychopharmacol* 2015; 35: 486–489.

[79] Ghanizadeh A, Mohammadi MR, Bahraini S, et al. Efficacy of N-Acetylcysteine Augmentation on Obsessive Compulsive Disorder: A Multicenter Randomized Double Blind Placebo Controlled Clinical Trial. *Iran J Psychiatry* 2017; 12: 2: 134-141.

[80] Li F, Welling MC, Johnson JA, Coughlin C, Mulqueen J, Jakubovski E, Coury S, Landeros-Weisenberger A, Bloch MH. N-Acetylcysteine for Pediatric Obsessive-Compulsive Disorder: A Small Pilot Study. *J*

Child Adolesc Psychopharmacol 2020; 30(1): 32-37. doi: 10.1089/cap.2019.0041.

[81] Witthauer C, Gloster AT, Meyer AH, Lieb R. Physical diseases among persons with obsessive compulsive symptoms and disorder: a general population study. *Soc Psychiatry Psychiatr Epidemiol* 2014; 49:2013–2022.

[82] Cuomo A, Maina G, Bolognesi S, Rosso G, Beccarini Crescenzi B, Zanobini F, Goracci A, Facchi E, Favaretto E, Baldini I, Santucci A, Fagiolini A. Prevalence and Correlates of Vitamin D Deficiency in a Sample of 290 Inpatients With Mental Illness. *Front Psychiatry* 2019; 10: 167. doi: 10.3389/fpsyt.2019.00167.

[83] Bokor G, Anderson PD. Ketamine: an update on its abuse. *J Pharm Pract* 2014; 27: 582–586.

[84] Chaki S. Beyond Ketamine: New Approaches to the Development of Safer Antidepressants. *Curr Neuropharmacol* 2017; 15(7): 963–976.

[85] Kompoliti K, Fan W, Leurgans S. Complementary and alternative medicine use in Gilles de la Tourette syndrome. *Mov Disord* 2009; 24: 2015-9.

In: Cysteine: Sources, Uses and Health Effects   ISBN: 978-1-53619-033-5
Editor: Taran Saunders   © 2021 Nova Science Publishers, Inc.

Chapter 3

COMPARATIVE STUDIES OF CYSTEINE RELEASE FROM DIFFERENT MESOPOROUS SILICA: MCM-41 AND MCM-48

Priyanka D. Solanki[1,2] and Anjali U. Patela[1,*]
[1]Department of Chemistry, Faculty of Science,
The Maharaja Sayajirao University, of Baroda,
Vadodara, Gujarat, India
[2]Department of Chemistry,
Government Science College Dhanpur,
Government of Gujarat, India

ABSTRACT

The present chapter describes the functionalization of MCM-41 and MCM-48 by an inorganic moiety, 12-tungstophophoric acid (TPA), Cysteine loading, characterization and *in vitro* release of Cysteine at body temperature under different conditions. A study on **Release** kinetics was carried out using First order release kinetic model while the mechanism was by Higuchi model. Further, to see the influence of **TPA** on release rate,

---

[*] Corresponding Author's E-mail: Anjali.patel-chem@msubaroda.ac.in.

release profile obtained from pure MCM-41 and MCM-48 were compared with functionalized materials i.e., TPA-MCM-41 and TPA-MCM-48. Finally, all data were correlated with geometry of the supports.

Keywords: MCM-41, MCM-48, 12-tungstophosphoric acid, cysteine, *in vitro* release, kinetics and mechanism

INTRODUCTION

Cysteine is a naturally occurring, sulfur-containing amino acid with a thiol group and is found in most proteins (Scheme 1). Its pro-drug, N-acetyl cysteine are used to treat Schizophrenia and reduce drug cravings [1]. It is a limiting substrate in the production of glutathione in the body. The reduction of intracellular levels of glutathione contributes to chronic inflammatory conditions, which are associated with cancer, neurogenerative, cardiovascular and infertility diseases resulting in high demand of Cysteine [2-4]. Current Cysteine therapies are administration of different derivatives such as N- cetyl-Cysteine. One of the major drawbacks of these therapies is high dosages that can provoke persistent damage and strong allergic reactions [5-8]. The mentioned problems can be overcome by encapsulating it in a carrier which can deliver it in controlled manner as well as protect it from enzymatic degradation.

Recently, mesoporous silica has gain tremendous attraction in this field for its significant properties such as ordered mesopores, high surface area, and well defined structure, presence of surface silanol group which can make them suitable as carrier. In 2009, Victor S-Y Lin and his group have reported *in vitro* release of cysteine from mesoporous silica nanoparticle [9]. In 2017, our group have reported *in vitro* release of cysteine from Mesoporous silica (MCM-41 and MCM-48) and study the effect of geometry of carrier on release rate of cysteine [10].

These studies encourage us to do further work in same direction. So as an extension of our work, here, we are reporting functionalization of MCM-41 and MCM-48 using inorganic moiety, 12-tungstophosphoric acid (TPA) and its use as carrier for cysteine.

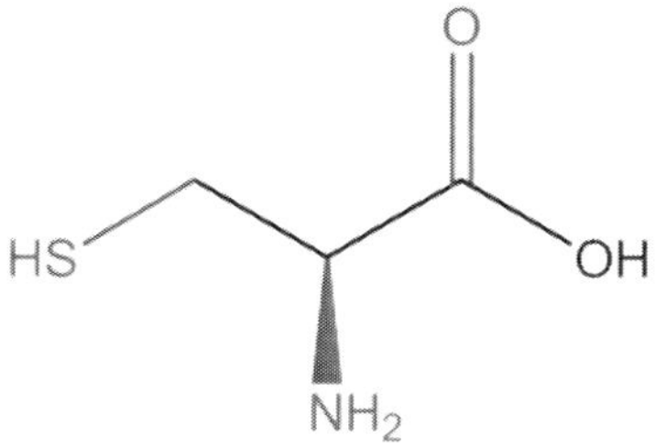

Scheme 1. Molecular structure of Cysteine.

The present chapter describes Functionalization of MCM-41 as well as MCM-48 by TPA, encapsulation of Cysteine into functionalized carriers and their characterization using various physicochemical techniques. *In vitro* release of cysteine was carried out at body temperature under static and dynamic condition. In order to see the influence of TPA on release rate, Cysteine was loaded on pure carrier and its release study was carried out under same experimental condition. To see the effect of geometry on release rate, release profiles obtained from both carriers were compared. Cysteine release kinetics and mechanism was also investigated using First order release kinetic model and Higuchi model.

EXPERIMENTAL

Materials

All chemicals used were of A. R. grade. 12-Tungstophosphoric acid (TPA), sodium hydroxide, Cetyl trimethyl ammonium bromide (CTAB), Tetraethyl orthosilicate (TEOS), n-Hexane, Mesitylene, Ethanol and Sodium hydroxide (NaOH) were used as received from Merck. Ninhydrin and Cysteine were used as received from Sigma Aldrich.

Simulated body fluid (SBF): NaCl (7.996 g), $NaHCO_3$ (0.350 g), KCl (0.224 g), $K_2HPO_4.3H_2O$ (0.22 g), $MgCl_2.6H_2O$ (0.305 g), 1 $molL^{-1}$ HCl (40 ml), $CaCl_2$ (0.278 g), Na_2SO_4 (0.071 g) and $NH_2C(CH_2OH)_3$ (6.057 g) were dissolved in small amount of water and then dilute to 1 L with distilled water [11, 12].

Synthesis of MCM-41

MCM-41 was synthesized using reported procedure [13] with modification. Surfactant (CTAB, 4.38 g) and NaOH (1.2 g) were dissolved in 200 mL double distilled water. When the solution became homogeneous, TEOS (20.8 g) was added quickly with stirring. After 5 min, mesitylene (8.64 g) and hexane (3.1 g) was added to the stirred mixture. The resulting thick mixture was stirred vigorously for 10 min and then heated at 85°C for 2 days with stirring. The resulting product was filtered, washed with double distilled water, dried at 100°C temperature. The obtained material was calcined at 550°C in air for 5 h and designated as MCM-41.

Synthesis of MCM-48

The synthesis of MCM-48 was carried out as reported in literature [14]. 2.4 g of CTAB was dissolved in 50 ml distilled water. To this, 50 ml ethanol (0.87 mole) and 12.6 ml ammonia (32 wt%, 0.2 mole) were added. The mixture was then stir for 10 min. when solution become homogenous, 3.4 g TEOS was added. After stirring for 2 h the resulting white solid was filtered and wash with distilled water. The template was removed by calcination for 823 K for 6 h. The resulting material was designated as MCM-48.

Functionalization of MCM-41 and MCM-48 Using 12-Tungstophosphoric Acid (TPA)

MCM-41 was functionalized using TPA by incipient wet impregnation method. 30% of TPA anchored to MCM-41 was synthesized. 1 g of MCM-41 was impregnated with an aqueous solution of TPA (0.3/30 g/mL of distilled water) and dried at 100 °C for 10 h. The obtained material was designated as TPA-MCM-41. By using similar method, functionalization of MCM-48 was also carried out and obtained materials were designated as TPA-MCM-48.

Leaching Test

Whether TPA truly acts as anchoring agent or not, leaching of TPA was investigated. TPA can be qualitatively characterized by the formation of heteropoly blue colour, when treated with a mild reducing agent such as ascorbic acid [15]. In the present study, this method was used for determining the leaching of TPA from the carrier. 1 g of TPA-MCM-41 was stirred in 10 mL water-ethanol mixture (40:60) for 24 h. Then 1 mL of the supernatant solution was treated with 10% ascorbic acid. Development of blue colour was not observed, indicating absence of any leaching. The same procedure was repeated with SBF as well SGF, in order to check the presence of any leached TPA. The absence of blue colour indicated no leaching of TPA. Similar method was also apply to TPA-MCM-48.

Loading of Cysteine into TPA-MCM-41 and TPA-MCM-48

Loading of Cysteine into TPA-MCM-41 was carried out by soaking method. 221 mg of TPA-MCM-41 was soaked in 10 mL solution of Cysteine (221 mg) in distilled water. The mixture is stirred in sealed vials to prevent the evaporation of solvent for 24 h. Then resulting mixture was filtered and washes with 10 mL of distilled water and dried at room temperature. The obtained material is designated as Cys/TPA-MCM-41. The loading amount of Cysteine was obtained by analyzing the filtrate using UV–Vis spectroscopy as well as by thermal analysis which shows 8.8% loading of Cysteine into the TPA-MCM-41. This was also calculated by TGA analysis of Cys/TPA-MCM-41. By using same method cysteine was also loaded into TPA-MCM-48 and the obtained materials were designated as Cys/TPA-MCM-48 and amount of cysteine lodaded into TPA-MCM-48 was found to be 8.0%.

Characterizations

^{31}P MAS NMR spectra were recorded by BRUKER Avance DSX-300, at 121.49 MHz using a 7 mm rotor probe with 85% phosphoric acid as an external standard. The spinning rate was 5–7 kHz. Catalyst samples, after treatment were kept in a desiccator over P_2O_5 until the NMR measurement. The ^{29}Si NMR spectra were recorded at Mercury Plus 300 MHz using a 5 mm Dual Broad Band rotor probe with Trimethyl silane (TMS) as an external standard. The FTIR spectra of all materials were obtained by using KBr palate on Perkin Elmer instrument. TGA of the materials were carried out using a Mettler Toledo Star SW 7.01 instrument under nitrogen atmosphere from 30 to 570 °C at the heating rate of 10 °C/min. Adsorption–desorption isotherms of all materials were recorded on a Micromeritics ASAP 2010 Surface area analyzer at liquid nitrogen temperature. From the adsorption desorption isotherms, specific surface area was calculated using BET method. The XRD pattern was obtained on PHILIPS PW-1830, with Cu Kα radiation (1.54 Å) and scanning angle from 0° to 10°. TEM analysis was carried out on JEOL (JAPAN) TEM instrument (model-JEM 100CX II) with accelerating voltage of 200 kV. The samples were dispersed in ethanol and ultrasonicated for 5–10 min. A small drop of the sample was then taken in a carbon coated copper grid and dried before viewing.

In Vitro Release Study of Cysteine

The release profile was obtained by soaking cysteine loaded materials in 150 ml of a SBF (1 mg of the Cysteine sample per ml of fluid). This release fluid was analyzed for Cysteine content by treating it with 10% ninhydrin solution at 570 nm. All the experiments were repeated three times.

RESULT AND DISCUSSIONS

Materials Characterizations

EDS analysis for TPA-MCM-41 and TPA-MCM-48 are shown in Table 1. The results obtained from EDS were in good agreement with the theoretical values.

^{29}Si and ^{31}P MAS NMR

Figure 1A shows the ^{29}Si MAS NMR spectra of the MCM-41 and TPA-MCM-41. A broad peak for MCM-41 between -90 and -125 ppm was observed which can be attributed to three main components with chemical shifts at -93, -103, and -110 ppm (Table 2, Figure 1A). These signals resulted from Q^2 (-93 ppm), Q^3 (-103 ppm), and Q^4 (-110 ppm) silicon nuclei, where Q^x corresponds to a silicon nuclei with x siloxane linkages, i.e., Q^2 to disilanol $Si-(O-Si)_2(-O-X)_2$, where X is H, Q^3 to silanol $(X-O)-Si-(O-Si)_3$, and Q^4 to $Si-(O-Si)_4$ in the framework.

For TPA-MCM-41, all the three bands were observed with broadening and slight shifting which also confirm that the structure of MCM-41 remains intact. A significant shift in Q^2, Q^3 and Q^4 bands are observed which confirm the interaction between surface Si-OH groups to TPA molecules.

Table 1. Results of elemental analysis in wt%

Materials	Elemental analysis (weight %)			
	W		P	
	Theoretical	By EDS	Theoretical	By EDS
TPA-MCM-41	19.0	18.0	0.32	0.30
TPA-MCM-48	19.0	18.3	0.32	0.31

Table 2. ^{29}Si chemical shift of MCM-41 and TPA-MCM-41

Materials	^{29}Si MAS NMR data		
	Q^2 ppm	Q^3 ppm	Q^4 ppm
MCM-41	-93	-103	-110
TPA-MCM-41	-89.39	-99.17	-108
MCM-48	-92	-99	-108
TPA-MCM-48	-88	-98	-108

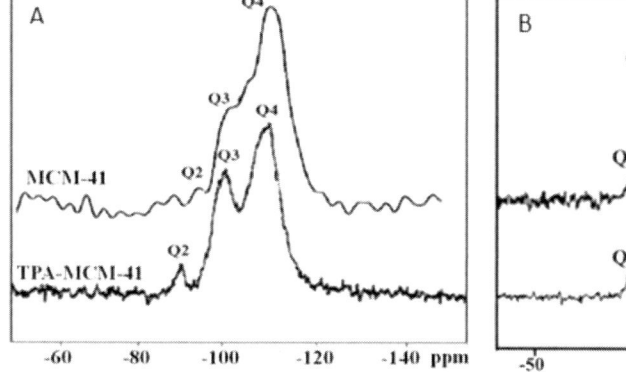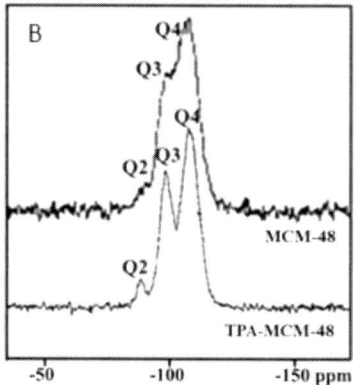

Figure 1. ^{29}Si MAS NMR of MCM-41, TPA-MCM-41, MCM-48 and TPA-MCM-48.

Figure 1B represents the ^{29}Si MAS NMR spectra of MCM-48 and TPA-MCM-48. A broad peak of MCM-48 between -90 to -125 ppm observed which also can be attributed to three main part of the peak with chemical shift at -92, -99 and -108 ppm (Figure 1B, Table 2). These signals resulted from Q^2 (-92 ppm), Q^3 (-99 ppm) and Q^4 (-108 ppm) silicon nuclei.

All the three, Q^2, Q^3 and Q^4 bands are observed in NMR spectra of TPA-MCM-48 (Figure 1b) suggesting the intact structure of MCM-48 even after functionalization. However, significant shift is observed in Q^2 band from -92 to -88 (Table 2) suggests the interaction of Si-OH group of MCM-48 with TPA.

^{31}P NMR is the most important method to study chemical environment around the phosphorus in heteropoly compounds. The ^{31}P NMR spectra of TPA, TPA-MCM-41 and TPA-MCM-48 are shown in Figure 2. Pure TPA shows single peak at -15.62 ppm and is in good agreement with the reported one [16]. The ^{31}P NMR spectra of TPA-MCM-41 and TPA-MCM-48 shows single peak at -12.97 ppm and 12.272 ppm respectively. The observed shift in NMR peak is attributed to the strong interaction of carrier with that of TPA as well as the presence of TPA inside the channels of carrier.

TGA

TGA of pure TPA, TPA-MCM-41, Cys/TPA-MCM-41, TPA-MCM-48, Cys/TPA-MCM-48 are shown in Figure 3. TPA exhibits weight loss in three stages at 100, 200 and 485°C. These can be attributed to initial weight due to adsorbed water, second weight loss due to loss of water of crystallization near 200°C to give the Keggin structure, which is stable on heating up to 350°C. The weight loss at 485°C may be attributed to the decomposition of the Keggin structure of TPA into the simple oxides [17]. TGA of TPA-MCM-41 shows initial weight loss of 3.6% due to the loss of adsorbed water. Second weight loss of 1.2% between 150 and 250°C corresponds to the loss of water of crystallization of Keggin ion. Further a gradual weight loss was also observed from 250 to 500°C due to the difficulty in removal of water contained in TPA molecules inside the channels of MCM-41 [17]. This type of inclusion causes the stabilization of TPA molecules inside the channels of MCM-41. TGA of Cys/TPA-MCM-41 shows initial weight loss of 2.3% due to the loss of adsorbed water. Further weight loss of 8.8%, from 200 to 450°C indicates the removal Cysteine which also confirms the 8.8% loading of Cysteine.

TGA curve of TPA-MCM-48 shows initial weight loss of 10-13% up to 150°C which indicate presence of adsorbed water. Second weight loss of 1-2% between 200 and 300°C corresponds to the loss of water of crystallization of Keggin ion. A gradual weight loss from 300 to 500°C was observed due to the difficulty in removal of water present in TPA molecules inside the channels of MCM-48. This type of inclusion can cause the stabilization of TPA molecules inside the channels of MCM-48. TGA curve of Cys/TPA-MCM-48 also shows initial weight loss of 2.3% up to 150°C respectively which is corresponding to removal of adsorbed water. Further weight loss of 8% from 200°C to 550°C may be due to the removal of cysteine from TPA-MCM-48. The amount of cysteine encapsulated into TPA-MCM-41 and TPA-MCM-48 are shown in Table 3.

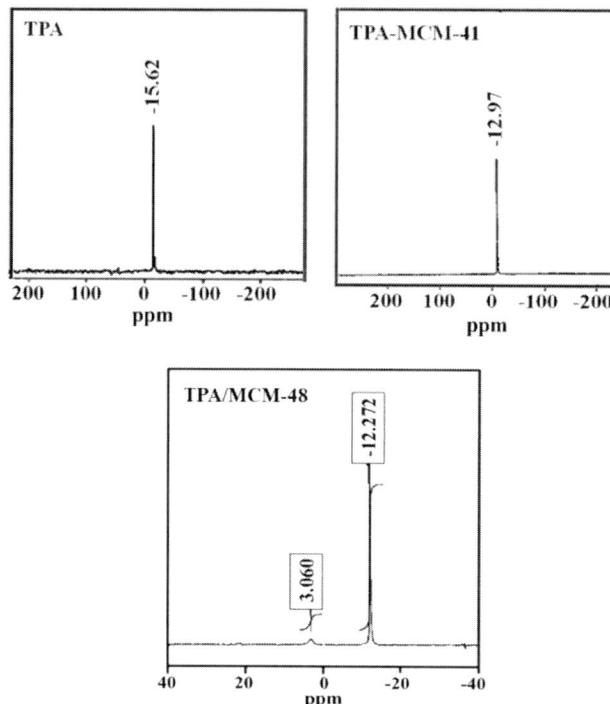

Figure 2. ^{31}P NMR spectra of pure TPA, TPA-MCM-41 and TPA-MCM-48.

Table 3. Amount of cysteine loaded

	% Loading	Amount of amino acids/drug encapsulated (mg/g of carrier)
MCM-41	15 ± 0.2	150 ± 2
TPA-MCM-41	8.8 ± 0.2	88 ± 2
MCM-48	14 ± 0.2	140 ± 2
TPA-MCM-48	8 ± 0.2	80 ± 2

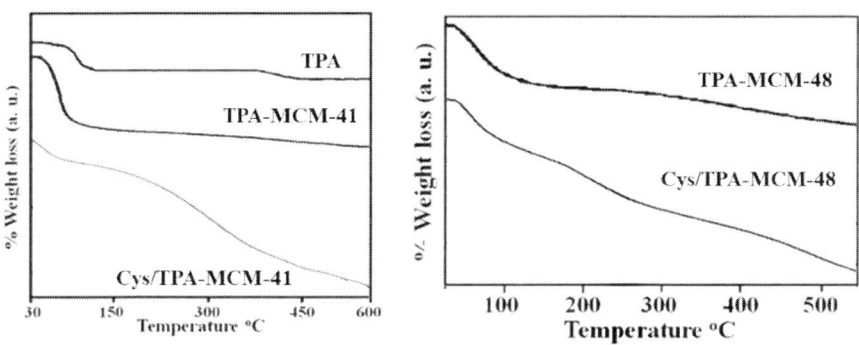

Figure 3. TGA of TPA, TPA-MCM-41, Cys/TPA-MCM-41, TPA-MCM-48, Cys/TPA-MCM-48.

FTIR

FTIR bands of MCM-41, TPA-MCM-41, Cys/TPA-MCM-41, MCM-48, TPA-MCM-48 and Cys/TPA-MCM-48 are shown in Figure 4. FTIR bands of MCM-41 shows broad band at 1100–1300 cm^{-1}, 3448 cm^{-1} corresponds to asymmetric stretching vibration of Si–O–Si and symmetric stretching vibration of Si–OH group, respectively. The bands at 801 and 498 cm^{-1} represent the symmetric stretching and bending vibration of Si–O–Si. The reported bands for TPA, at 1088, 987, and 897 cm^{-1} corresponding to P-O stretching, W–O symmetric stretching and W–O-W bending respectively, are absent in TPA-MCM-41. If TPA is dispersed onto the surface of MCM-41, the mentioned bands for TPA should be seen in the FT-IR spectra. The absence of respective FTIR bands of TPA in TPA-MCM-41 may due to the overlapping of TPA bands with that of MCM-41. The FTIR of Cys/TPA-MCM-41 shows entire bands related to TPA-MCM-41. Along with this, it shows additional bands at 1580 and 1404 cm^{-1} corresponding to CH$_2$ stretching vibration. The significant shifting in band of SH group from 2551 to 2380 cm^{-1} indicates the interaction of Cysteine through SH group.

The FTIR of MCM-48 shows a broad band around 1100 and 1165 cm^{-1} corresponding to asymmetric stretching of Si-O-Si. The bands at 640 cm^{-1} and 458 cm^{-1} are due to symmetric vibrations of Si-O-Si and bending vibrations of Si-O, respectively. FTIR spectrum of Cys/TPA-MCM-48 also shows whole bands related to TPA-MCM-48 suggesting the intact structure of TPA-MCM-48. The significant shifting in band of SH group from 2551 to 2600 cm^{-1} indicates the interaction of Cysteine through SH group.

Nitrogen Adsorption-Desorption Isotherm

Figure 5 shows Nitrogen adsorption-desorption isotherm of MCM-41, TPA-MCM-41, Cys/TPA-MCM-41, MCM-48, TPA-MCM-48 as well as Cys/TPA-MCM-48 and textural properties of these are shown in Table 4. The isotherm is type (IV) in nature according to the IUPAC classification and exhibited H1 hysteresis loop which is a characteristic of mesoporous solids for all the systems. Decrease in all structural parameters of cysteine loaded TPA-MCM-41 as well as TPA-MCM-48 suggests the insertion into the porous channels of TPA-MCM-41 and TPA-MCM-48 respectively. In all the case, the basic structure of MCM-41 and MCM-48 remains intact. Further, functionalization by TPA does not alter the structure of carrier.

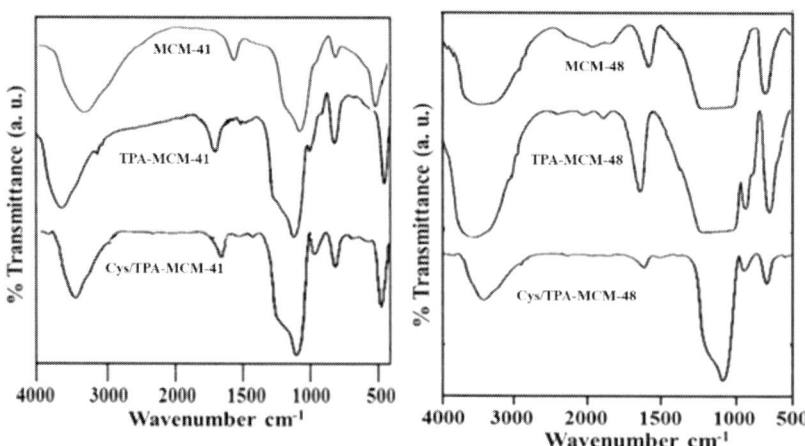

Figure 4. FTIR spectra of MCM-41, TPA-MCM-41, Cys/TPA-MCM-41, MCM-48, TPA-MCM-48, Cys/TPA-MCM-48.

Table 4. Textural properties of materials and amount of drug encapsulated into TPA-MCM-41

Materials	Specific surface area (m²/g)	Pore volume (cm³/g)
MCM-41	890	1.19
TPA-MCM-41	622	0.67
Cys/TPA-MCM-41	284	0.35
MCM-48	1141	0.67
TPA-MCM-48	566	0.22
Cys/TPA-MCM-48	222	0.210

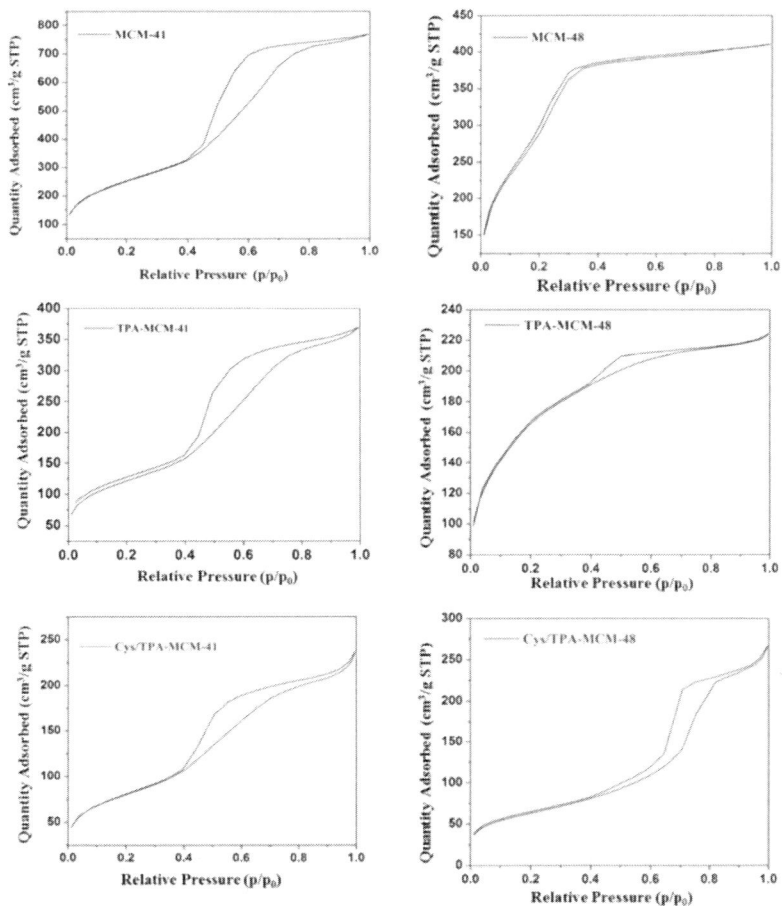

Figure 5. Nitrogen adsorption-desorption isotherms of all materials.

Low Angle Powder XRD

Figure 6 shows low angle powder XRD of MCM-41, TPA-MCM-41, Cys/TPA-MCM-41, MCM-48, TPA-MCM-48 and Cys/TPA-MCM-48. The low angle XRD of MCM-41 displays an intense diffraction peak at $2\theta = 2°$ which are assigned to the lattice faces (100), suggesting a hexagonal symmetry of MCM-41. In addition to this weak peaks are observed at $2\theta = 3\text{-}5°$ with very low intensity which are sometimes difficult to observed due to instrumental error [18]. Further, Lu et al. have reported similar pattern of XRD peaks of MCM-41 [19]. It is well known that the low angle XRD pattern are sensitive to pore filling and loaded materials show lowered intensity of characteristic peak compared to pure one. It is well known that the low angle XRD pattern are sensitive to pore filling and loaded materials show lowered intensity of characteristic peak compared to pure one and this reflect in the XRD pattern of all the materials. XRD pattern of TPA-MCM-41 and Cys/TPA-MCM-41 shows characteristics peak at $2\theta = 2°$ with lower intensity. Absence of any other peak indicates the insertion as well as homogeneous distribution of Cysteine into the mesoporous channels of TPA-MCM-41.

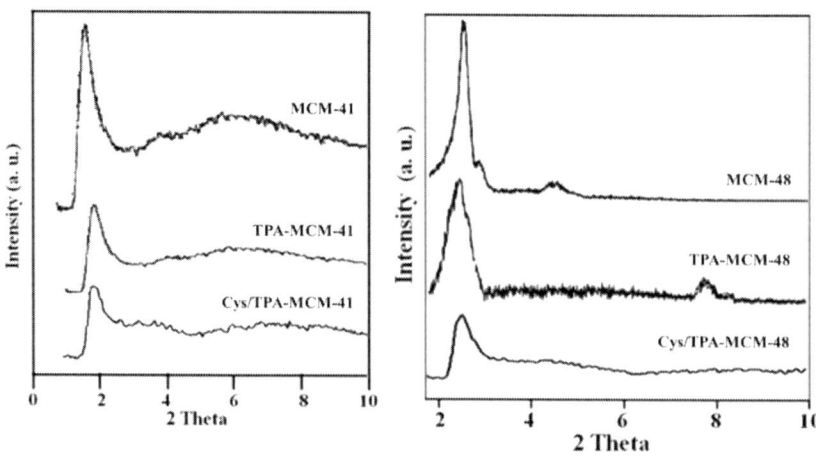

Figure 6. Low angle XRD of MCM-41, TPA-MCM-41, Cys/TPA-MCM-41, MCM-48, TPA-MCM-48, Cys/TPA-MCM-48.

The XRD pattern of MCM-48 shows main characteristic peaks at 2.3° which indicates the presence of intact structure of MCM-48. It also shows broad shoulder at 3.0° corresponding to a plane 211 and 220, respectively. Several peaks in the range of 4-5° diffraction angles correspond to the reflections of 400, 321 and 420 planes of a typical MCM-48 meso-structure with Ia3d cubic symmetry. XRD pattern of TPA-MCM-48 and Cys/TPA-MCM-48 also shows same peak with lower intensity. Absence of any other peaks indicated the well-ordered distribution of cysteine into TPA-MCM-48.

TEM

Figure 7 Shows TEM images of MCM-41, TPA-MCM-41, Cys/TPA-MCM-41, MCM-48, TPA-MCM-48 and Cys/TPA-MCM-48 at different magnifications. TEM images of MCM-41 shows hexagonal and very well ordered porous structure. TEM images of TPA-MCM-41 shows absence of crystalline phase of TPA inside the channels of MCM-41 indicating high dispersion of TPA and absence of agglomeration. TEM images of CYs/TPA-MCM-41 shows ordered porous structure with absence of any agglomeration indicating the well dispersion of amino acids/drug into TPA-MCM-41. TEM images of MCM-48, TPA-MCM-48 and Cys/TPA-MCM-48 shows spherical morphology. Further, functionalization as well as loading of Cysteine does not alter the structure of MCM-48. It also shows absence of any agglomeration indicating well-ordered homogenous dispersion of cysteine.

In Vitro Release Study

Effect of Stirring on Release Rate

In order to evaluate the effect of stirring on release rate of Cysteine, *in-vitro* release of Cysteine was carried out under stirring condition as well as static condition and results are shown in Figure 8. Both the systems,

Cys/TPA-MCM-41 (Figure 8A) and Cys/TPA-MCM-48 (Figure 8B) show slower release under static condition. For Cys/TPA-MCM-41, initially, 28% and 34% of Cysteine is released under static and stirring condition respectively. After that, release of Cysteine in controlled manner is observed in stirring condition and reached to maximum at 30 h. Similar observation is obtained for Cys/TPA-MCM-48. Initially, 25% and 28% of Cysteine is released under static and stirring condition respectively. After that, release of Cysteine, in controlled manner is observed in stirring condition and reached to maximum at 30 h. The observed slower release of drug under static condition is may be due to the slower diffusion of drug which eventually decreases the dissolution rate of Cysteine from materials to the release medium and thus requires high time for complete release of Cysteine. Here more convenient result is obtained under stirring condition.

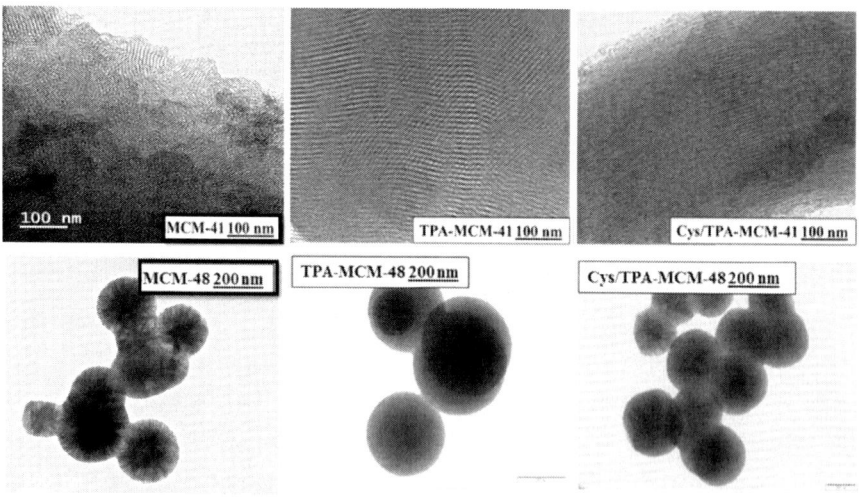

Figure 7. TEM images of all the materials at different magnification.

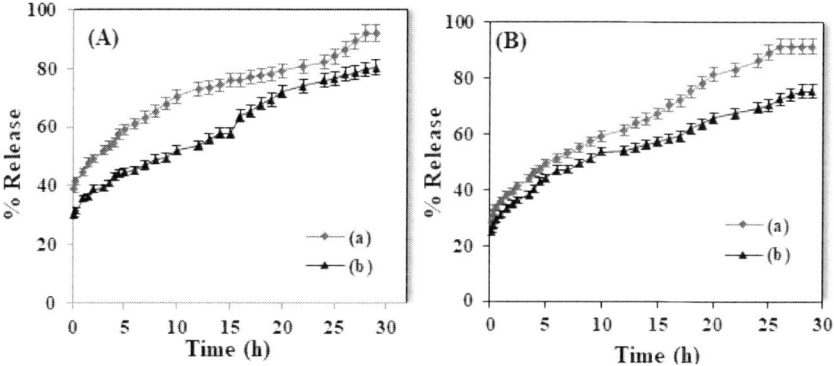

Figure 8. *In vitro* release profile of **(A)** Cys/TPA-MCM-41 and **(B)** Cys/TPA-MCM-48 under (a) stirring and static (b) condition.

Effect of TPA on Release Rate

To see the effect of TPA on release rate of Cysteine, release profile of Cys/MCM-41 and Cys/TPA-MCM-41 were compared and results are shown in Figure 9A. Initially, 38% and 30% of Cys is released and reached to 72% and 61% up to 10 h from MCM-41 and TPA-MCM-41, respectively. It reached to 91% and 84% up to 30 h for MCM-41 and TPA-MCM-41, respectively. Here, slower release profile is obtained for Cys/TPA-MCM-41 as compared to Cys/MCM-41.

Similarly, release profile of Cys/TPA-MCM-48 and Cys/MCM-48 were compared and result is shown in Figure 9B. Initially, 34% and 28% of Cysteine is released and reached up to 76% and 59% in 10 h from MCM-48 and TPA-MCM-48, respectively. It reached up to 91% in 30 h for MCM-48 and TPA-MCM-48, respectively. Here, slower release profile is obtained for Cys/TPA-MCM-48 as compared to Cys/MCM-48.

The slower diffusion of cysteine is may be because of the more attractive interaction between the Cys molecules and TPA-MCM-41. As stated earlier, TPA has terminal free oxygen through which it can bind with drug. This is may be the reasons of slower release of Cys from functionalized carriers.

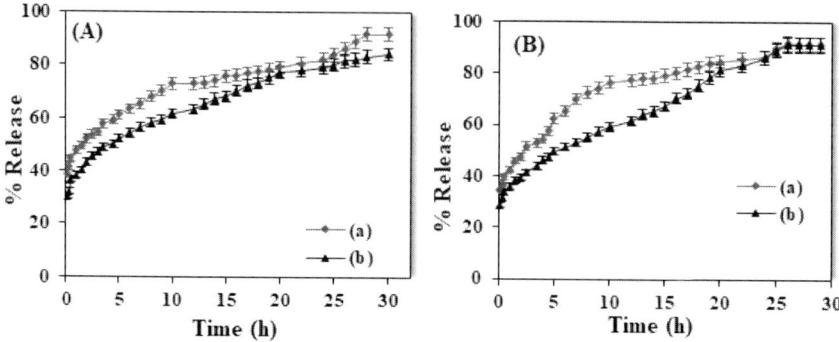

Figure 9. Comparison of release profile of **(A)** (a) Cys/MCM-41 with (b) Cys/TPA-MCM-41 and **(B)** (a) Cys/MCM-48 with (b) Cys/TPA-MCM-48.

Effect of Geometry of Carrier on Release Rate

To see the effect of geometry, release profile obtained from two different carriers have been compared and results are shown in Table 5. Results shows that more delayed and slower release was observed in case of pure MCM-41 as compared to MCM-48 (Figure 9). The obtained results are attributed to the well-massed transportation of the 3D interconnected pore system of MCM-48, which reduces the diffusion hindrance and assists drug diffusion into the medium. After functionalization, it was found that release rate became slower for both the systems as compared to pure carriers, showcasing the superiority of functionalized carriers. However, release rate in case of TPA-MCM-48 is slower as compared to TPA-MCM-41. This indicates that, though release rate is controlled to a large extent by functionalizing agent, the role of geometry of the carrier cannot be neglected (Table 5).

Table 5. Comparison of release profile of cysteine obtained from TPA-MCM-41 and TPA-MCM-48

Materials	% of Drug release		
	Initial	After 10h	30 h
Cys/TPA-MCM-41	30%	72%	91%
Cys/TPA-MCM-48	28%	59%	91%

Kinetics and Mechanism

In order to understand the release kinetics of Cysteine, the drug release data of all systems were Fitted with first order release kinetics model and Higuchi model [20-22].

The First order release model describes the release of drug which is concentration dependent and applied for water soluble drugs (Eq. 1).

$$\text{Log } C_0 - \text{Log } C_t = kt/2.303 \quad (1)$$

where C_0 is the initial concentration of drug and C_t is the concentration of drug at time t. k is the first order release kinetic constant. The plot of log of % remaining versus time will be linear with the negative slope for this model.

The Higuchi mechanism describes the percentage of release versus square root of time dependent process based on Fickian diffusion (Eq. 2). This model was used to describe the drug release from granular spherical partials and follow the dissolution mechanism.

$$Q = KHt^{1/2} \quad (2)$$

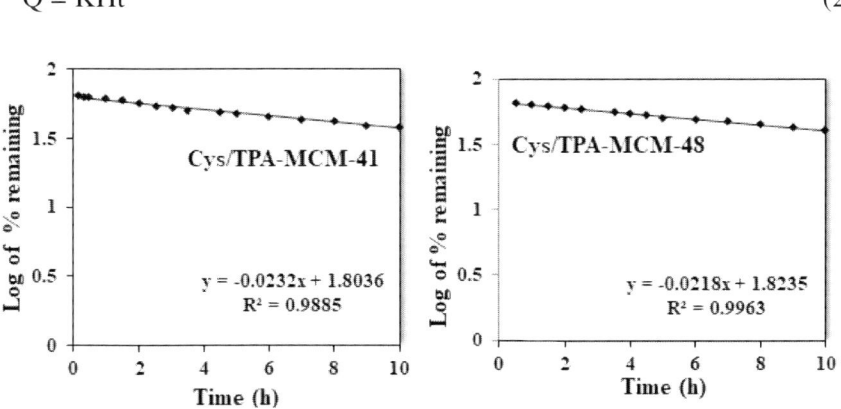

Figure 10. First order release kinetic model of Cys/TPA-MCM-41, and Cys/TPA-MCM-48.

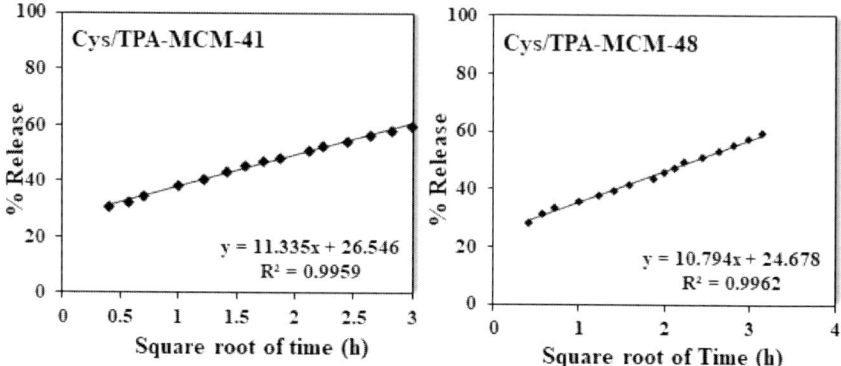

Figure 11. Higuchi model of Cys/TPAMCM-41 and Cys/TPA-MCM-48.

where, Q is the amount of drug release, t is the time and KH is the Higuchi constant. This model shows the mechanism of drug release which involves the simultaneous penetration of SBF into the pore, dissolution of drug molecule into the SBF and diffusion of drug molecule through the pore.

First Order Release Kinetic Model

In order to analyze Cysteine release in a detail and obtain the possible release mechanism, release data up to 10 h is fitted with First order release kinetic model as well as Higuchi model. Figure 10 shows the First order release kinetic model of Cys/TPA-MCM-41 and Cys/TPA-MCM-48 where Log of percentage remaining data is plotted against time. The release of Cysteine follows the first order kinetic with linearity and correlation coefficient (R^2) value for Cys/TPA-MCM-41 (0.9885) and Cys/TPA-MCM-48 (0.9963).

Higuchi Model

The Higuchi model (Figure 11) describes the percentage release versus square root of time dependent process based on Fickian diffusion. According to this model release mechanism of Cysteine involves simultaneous penetration of SBF into the pores, dissolution of drug molecule and diffusion of these molecules from the pores. The *in vitro* Cysteine release data is best fitted with Higuchi model. The release mechanism of Cysteine is best explained by this model with high linearity and high correlation co-efficient (R^2) value for Cys/TPA-MCM-41(0.9959) and Cys/TPA-MCM-48 (0.9962).

CONCLUSION

In this chapter first time we are reporting comparisons of release profile of cysteine from functionalized MCM-41 and MCM-48. *In vitro* release studies shows that more controlled and delayed release was obtained with TPA-MCM-48 system suggesting its superiority. Because of having 3D interconnected porous structure in case of MCM-48, fine mass transportation can occur. However, presence of TPA delayed the release rate of cysteine and hence effect of TPA became dominant over effect of geometry of carrier which was reflected in release rate. Further kinetics and mechanism study shows that release of cysteine follows first order kinetics with Fickian diffusion mechanism.

ACKNOWLEDGMENTS

Authors are thankful to Department of Chemistry, Faculty of Science, The Maharaja Sayajirao University of Baroda for BET and TGA analysis.

REFERENCES

[1] Cook, J., Baker, A. & Yin, W. U S Patent application 20090281109.
[2] Senthil, K., Aranganathan, S. & Nalini, N. (2004). Evidence of oxidative stress in the circulation of ovarian cancer patients. *Clin Chim Acta, 339,* 27-32.

[3] Gilgun-Sherki, Y., Melamed, E. & Offen, D. (2001). Oxidative stress induced-neurodegenerative diseases: the need for antioxidants that penetrate the blood brain barrier. *Neuropharmacology*, *40*, 959-975.

[4] Black, P. H. & Garbutt, L. D. (2002). Stress, inflammation and cardiovascular disease. *J. Psychosom. Res*, *52*, 1-23.

[5] Anderson, M. E. & Meister, A. (1987). Intracellular delivery of cysteine. *Methods Enzymol*, *143*, 313-325.

[6] Santangelo, F. (2003). Intracellular thiol concentration modulating inflammatory response: influence on the regulation of cell functions through cysteine prodrug approach. *Curr. Med. Chem*, *10*, 2599-2610, (2003).

[7] Brack, C., Labuhn, M. & Bechter-Thüring, E. (1997). N-Acetylcysteine slows down ageing and increases the life span of Drosophila melanogaster. *Cell. Mol. Life. Sci*, *53*, 960.

[8] Auzinger, G. & Wendon, J. (2008). Intensive care management of acute liver failure. *Curr Opin Crit Care*, *14*, 179-188.

[9] Mortera, R., Juan, V. E., BIgor, I. S., Garrone, E., Onida, B. & Lin, SY. V. (2009). Cell-induced intracellular controlled release of membrane impermeable cysteine from a mesoporous silica nanoparticle-based drug delivery system. *Chem. Commun.*, 3219–3221.

[10] Pathan, S., Solanki, P. & Patel, A. (2017). Cysteine and N-acetyl cysteine encapsulated mesoporous silica: synthesis, characterization and influence of parameters on *in vitro* controlled release. *J Porous Mater.*, *24*, 1105–1115.

[11] Tanga, Q., Xu, Y., Wu, D. & Sun, Y. (2006). A study of carboxylic-modified mesoporous silica in controlled delivery for drug famotidine. *Journal of Solid State Chemistry.*, *179*, 1513-1520.

[12] Yamamuro, T. J. (1990). Solutions able to reproduce *in vivo* surface-structure changes in bioactive glass-ceramic A-W3. *Biomedical Materials Research.*, *24*, 721-734.

[13] Myong, H. & SteinLim, A. (1999). Comparative Studies of Grafting and Direct Syntheses of Inorganic−Organic Hybrid Mesoporous Materials. *Chem. Mater.*, *11*, 3285–3295.

[14] Kumar, D., Schumacher, K., Hohenesche, C. D. F. V., Grun, M. & Unger, K. K. (2001). MCM-41, MCM-48 and related mesoporous adsorbents: their synthesis and characterization. *Colloid Surf. A*, 187–188, 109–116.
[15] Yadav, G. D. & Bokade, V. V. (1996). Novelties of heteropoly acid supported on clay: etherification of phenethyl alcohol with alkanols. *Appl. Catal. A*, *147*, 299-323.
[16] Okuhara, T., Mizuno, N. & Misono, M. (1996). *Adv. Catal.*, *41*, 113.
[17] Kamalakar, G., Komura, K., Kubota, Y. & Sugi, Y. (2006). Friedel–Crafts benzylation of aromatics with benzyl alcohols catalyzed by heteropoly acids supported on mesoporous silica. *J. Chem. Technol. Biotechnol.*, *81*, 981-988.
[18] Kruk, M. & Jaroniec, M. (1999). Characterization of Highly Ordered MCM-41 Silicas Using X-ray Diffraction and Nitrogen Adsorption. *Langmuir*, *15*, 5279-5284.
[19] Zhu, H. Y., Zhao, X. S., G. Q. Lu, G. Q. & Do, D. D. (1996). Improved Comparison Plot Method for Pore Structure Characterization of MCM-41. *Langmuir*, *12*, 6513-6517.
[20] Costa, P. & Sousa Lobo, J. M. (2001). Modeling and comparison of dissolution profiles. *Eur. J. Pharm. Sci.*, *13*, 123–133.
[21] Singhvi, G. & Singh, M. (2001). *Int. J. Pharm. Stud. Res.*, *2*, 1, 77–84.
[22] Salome, C., Godswill, O. & Ikechukwu, O. (2013). Kinetics and mechanisms of drug release from swellable and non swellable matrices: A review. *Res. J. Pharm. Biol. Chem. Sci.*, *4*, 2, 97–103 (2013).

BIOGRAPHICAL SKETCHES

Priyanka Solanki

Affiliation: Department OF Chemistry, Government Science College, Dhanpur, Dahod, Gujarat

Education: PhD (Chemistry)

Research and Professional Experience: 6 years (during PhD)

Professional Appointments: 31st July 2018 as assistant Professor

Honors: NET JRF 2012

Publications from the Last 3 Years:

[1] Solanki, P; Patel, A. Encapsulation of Aspirin into parent and functionalized MCM-41, *in vitro* release as well as kinetics, *J. Porous Mater.*, 26, 1523-1532, 2019.

[2] Solanki, P; Patel, S; Devkar, R; Patel, A. Camptothecin encapsulated into functionalized MCM-41: *In vitro* release study, cytotoxicity and kinetics, *Mater. Sci. Eng. C*, 98, 1014, 2019.

[3] Solanki, P; Patel, A. L-Arginine encapsulated mesoporous MCM-41 nanoparticles: A study on *in-vitro* release as well as kinetics, *Adv. Porous Mater.*, 6, 80, 2018.

[4] Solanki, P; Patel, A. *In-vitro* release of L-arginine and cysteine from MCM-48: A study of effect of size of active biomolecules on release rate, *J. Porous Mater.*, DOI: 10.1007/s10934-018-0561-z, 2018.

[5] Pathan, S; Solanki, P; Patel, A. Functionalized SBA-15 for controlled release of poorly soluble drug, Erythromycin, *Microporous Mesoporous Mater.*, 258, 114, 2018.

[6] Pathan, S; Solanki, P; Patel, A. Cysteine and N-acetyl cysteine encapsulated mesoporous silica: synthesis, characterization and influence of parameters on *in vitro* controlled release, *J. Porous Mater.*, 24, 1105, 2017.

Anjali Patel

Affiliation: Department OF Chemistry, Faculty of Science, The Maharaja Sayajirao University of Baroda

Education: PhD (Chemistry)

Research and Professional Experience: Twenty six years (after PhD)

Professional Appointments: Professor and Head, Chemistry Department, The M. S. University of Baroda

Honors: Fellow of Royal Society of Chemistry (FRSC)

Publications from the Last 3 Years:

Reviews:

[1] Sadasivan, R; Patel, A. Unmodified and modified copper polyoxometalates as catalysts for oxidation of alkenes: Kinetic and mechanistic investigation, *Inorganica Chimica Acta*, 510, 119757, 2020.
[2] Patel, A; Narkhede, N; Patel, A. Anchored Silicotungstates: Effect of Supports on Catalytic Activity, *Catal. Surv. Asia*, 23, 257-264, 2019.

Publications:

[1] Patel, A; Patel, A. Designing of Highly Active and Sustainable Encapsulated Stabilized Palladium Nanoclusters as well as Real Exploitation for Catalytic Hydrogenation in Water, *Catal Lett*, 10.1007/s10562-020-03327-4, (2020).
[2] Anjali Patel, Dhruvi Pithadia. Low temperature synthesis of bio-fuel additives via valorisation of glycerol with benzaldehyde as well as furfural over a novel sustainable catalyst, 12-tungstosilicic acid

anchored to ordered cubic nano-porous MCM-48, *Applied Catalysis A, General*, 602, 117729, 2020.

[3] Anjali Patel, Jay Patel. Nickel salt of phosphomolybdic acid as a bifunctional homogeneous recyclable catalyst for base free transformation of aldehyde into ester, *RSC Adv.*, 10, 22146, 2020.

[4] Patel, A; Patel, K. Polyoxometalate based hybrid chiral material: Synthesis, characterizations and aerobic asymmetric oxidation reaction, *J. Coord. Chem.*, 19-21, 3417-3429, 2019.

[5] Patel, A; Sadasivan, R. Hybrid Catalyst Based on Cu Substituted Phosphotungstate and Imidazole: Synthesis, Spectroscopic Characterization, Solvent Free Oxidation of Styrene with TBHP and Kinetics, *Catal. Lett.*, DOI: 10.1007/s10562-019-02979-1, 2019.

[6] Sadasivan, R; Patel, A. Flexible oxidation of styrene using TBHP over Zirconia supported mono-copper substituted phosphotungstate, *RSC Adv.*, 9, 27755-27767, 2019.

[7] Solanki, P; Patel, A. Encapsulation of Aspirin into parent and functionalized MCM-41, *in vitro* release as well as kinetics, *J. Porous Mater.*, 26, 1523-1532, 2019.

[8] Patel, A; Patel, A. Selective C=C Hydrogenation of Unsaturated Hydrocarbons in Neat Water Over Stabilized Palladium Nanoparticles Via Supported 12-Tungstophosphoric Acid, *Catal. Lett.*, 149, 1476, 2019.

[9] Solanki, P; Patel, S; Devkar, R; Patel, A. Camptothecin encapsulated into functionalized MCM-41: *In vitro* release study, cytotoxicity and kinetics, *Mater. Sci. Eng. C*, 98, 1014, 2019.

[10] Patel, A; Patel, A. Nickel exchanged supported 12-tungstophosphoric acid: synthesis, characterization and base free one-pot oxidative esterification of aldehyde and alcohol, *RSC Adv.*, 9, 1460, 2019.

[11] Patel, A; Patel, A; Narkhede, N. Hydrogenation of Cyclohexene in Aqueous Solvent Mixture Over a Sustainable Recyclable Catalyst Comprising Palladium and Monolacunary Silicotungstate Anchored to MCM-41, *Eur. J. Inorg. Chem*, 2019, (3-4), 423, 2019.

[12] Sadasivan, R; Patel, A; Ballabh, A. Investigation of catalytic properties of Cs salt of di-copper substituted phosphotungstate,

$Cs_7[PW_{10}Cu_2(H_2O)O_{38}]$ in epoxidation of styrene, *Inorg. Chim. Acta*, 487, 245, 2019.

[13] Patel, A; Sadasivan, R. Cs Salt of Undecatungstophospho(aqua) Cuprate(II): Microwave Synthesis, Characterization, Catalytic and Kinetic Study for Epoxidation of cis-Cyclooctene with TBHP, *ChemSelect*, 3, 11087, 2018.

[14] Patel, A; Patel, A. Stabilized Palladium Nanoparticles: Synthesis, Multi-spectroscopic Characterization and Application for Suzuki–Miyaura Reaction, *Catal. Lett.*, 148, 3534-3547, 2018.

[15] Solanki, P; Patel, A. L-Arginine encapsulated mesoporous MCM-41 nanoparticles: A study on *in-vitro* release as well as kinetics, *Adv. Porous Mater.*, 6, 80, 2018.

[16] Solanki, P; Patel, A. *In-vitro* release of L-arginine and cysteine from MCM-48: A study of effect of size of active biomolecules on release rate, *J. Porous Mater.*, DOI: 10.1007/s10934-018-0561-z, 2018.

[17] Pathan, S; Solanki, P; Patel, A. Functionalized SBA-15 for controlled release of poorly soluble drug, Erythromycin, *Microporous Mesoporous Mater.*, 258, 114, 2018.

[18] Singh, S; Patel, A. Value added products derived from Biodiesel waste Glycerol: activity, selectivity, kinetic and thermodynamic evaluation over anchored lacunary phosphotung state, *J. Porous Mater.*, 24, 1409, 2017.

[19] Patel, A; Sadasivan, R. Microwave assisted one pot synthesis and characterization of Cesium salt of di-copper substituted phosphotung state and its application in the selective epoxidation of cis-cyclooctene with tert-butyl hydroperoxide, *Inorg. Chim. Acta*, 458, 101, 2017.

[20] Patel, A. Functionalization of Keggin-type nickel substituted phosphotung state by imidazole: synthesis, characterization, and catalytic activity, *J. Mater. Sci.*, 52, 4689, 2017.

[21] Pathan, S; Solanki, P; Patel, A. Cysteine and N-acetyl cysteine encapsulated mesoporous silica: synthesis, characterization and influence of parameters on *in vitro* controlled release, *J. Porous Mater.*, 24, 1105, 2017.

In: Cysteine: Sources, Uses and Health Effects ISBN: 978-1-53619-033-5
Editor: Taran Saunders © 2021 Nova Science Publishers, Inc.

Chapter 4

OPTICAL SENSORS FOR CYSTEINE: A BRIEF OVERVIEW

Goutam K. Patra[*], *Amit K. Manna* and *Meman Sahu*
Department of Chemistry,
Guru Ghasidas Vishwavidyalaya, Bilaspur (C. G), India

ABSTRACT

Amino acids are crucially involved in an innumerable of biological processes. Any irregular changes in physiological level of amino acids often manifest in common metabolic disorders, serious neurological conditions and cardiovascular diseases. Among the amino acid series cysteine plays a major role in various physiological processes like protein synthesis, detoxification and metabolism of living organism. Its deficiency can cause several problems like hematopoiesis disease, retarded growth of children, skin lesion and loss of leucocyte etc. Therefore rapid and selective detection and quantification of cysteine in biological relevant samples has become very essential to its efficient clinical finding in recent years. In this book chapter our objective is to discuss the recent developments in designing the fluorescent and colorimetric sensors (optical sensors) for selective and sensitive detection of cysteine.

---

[*] Corresponding Author's E-mail: patra29in@yahoo.co.in.

Keywords: amino acids, cysteine, optical sensors, fluorimetry, colorimetry, supramolecules

INTRODUCTION

Cysteine, one of the important amino acid play important role in the biological system. Cys is a metabolic product of Hcy and a precursor of the antioxidant GSH, and its normal intracellular level remains to be 30–200μM [1]. The deficiency of Cys could cause edema, leucocyte loss, liver damage as well as neurotoxicity, whereas the excess levels of Cys might relate to cardiovascular and Alzheimer's diseases [2-4]. Hence, it has attracted intense interest in the development of novel strategy for detection and imaging of the intracellular Cys, which will further contribute to the better understanding the pathology of associated diseases and their early diagnosis and treatment. Modification in the cysteine moiety has great role in the biological function known as biological modification. The thiol group in the cysteine is the key component. The thiol group provides easy way of modification in the cysteine moiety ultimately in the protein.

The sulfur atom of cysteine (Cys) provides a considerable range of chemical reactivity and structural flexibility in the proteome (complete set of protein expressed by an organism). The presence of conserved Cys in motifs of proteins found in essentially all life forms indicates that these chemical characteristics were harnessed in early evolution to support enzyme catalysis, transcriptional regulation, protein folding, and three dimensional structures. A remarkable range of biologic functions is supported by Cys because sulfur is stable in multiple coordinate covalent bonds with the major atoms of living organisms (C, H, O, N, P), forms stable coordinate complexes with transition metal ions (Zn, Fe, Cu), and is stable at a range of oxidation states ($-SH$, $-SS-$, $-SO_2^{-1}$, $-SO_3^{-1}$). Additionally, Cys thiols differentially undergo reversible ionization to the negatively charged thiolate form over the physiologic range of pH to flexibly optimize the functions of specific peptidyl Cys.

Biothiols such as cysteine (Cys), homocysteine (Hcy), and glutathione (GSH) are essential biomolecules and participate in molecular and physiological processes in living systems in the maintenance of redox homeostasis, intracellular signal transduction, and human metabolism [5, 6]. Recently, they have received much research attention [7, 8]. Cys is an important amino acid that not only plays a pivotal role in many physiological processes including protein synthesis, detoxification, and metabolism but is also very closely related to many serious diseases such as slow growth in children, liver damage, muscle and fat loss, skin lesions, and weakness [9-11]. Therefore, the detection of Cys is very important for potential disease diagnosis. Various detection techniques such as fluorescence assays, high-performance liquid chromatography, capillary electrophoresis, immunoassays, and colorimetry [12, 13] have been developed. Fluorescence detection is particularly attractive because of its high selectivity, high sensitivity, low cost, and its great potential for bioimaging with fluorescent probes [14]. Distinguishing Cys from other biothiols, particularly Hcy and GSH, remains a challenge because of their similar structures and reactivity [15, 16]. Many fluorescent probes have been developed for the detection of Cys [17, 18], most of these Cys probes were based on fluorescence measurements at a single wavelength and some are ratiometric fluorescent probes allow the measurement of emission intensities at two different wavelengths, which provides a built-in correction of environmental effects and can also increase the dynamic range of the fluorescence measurement.

The design and construction of chemosensors with high selectivity and sensitivity for amino acids has received particular interest. Although a number of fluorescent probes for Cys detection and imaging have been reported, most of them could not discriminate Cys from Hcy/GSH due to their similar molecular structures and reactivity [19-22].

There is still much room for improvement in terms of rapidity, sensitivity, selectivity and cost-effectiveness to recognize those biologically important organic molecules such as amino acids. A molecule containing conjugated electron donor (D) and electron acceptor (A) usually undergoes an intramolecular charge transfer (ICT) upon electronic excitation [23]. The ICT and hence an elongation of the p electron conjugation occurring upon

Franck–Condon excitation contributes to the absorption profile. The process of ICT fluorescence plays a key role in the biological systems such as photosynthesis [24]. Therefore, it is very important to develop highly sensitive and selective assays for cysteine (Cys).

CHEMOSENSORS FOR CYSTEINE DETECTION

De Silva demonstrated the first use of PET sensors [25] for thiols in 1998, it was in 2004, Martı'nez-Ma'n˜ez and coworkers developed two squaraine based fluorescent chemosensors (1a) and (1b) for the detection of thiols [26].

Their solutions showed color changes from blue to colorless along with fluorescence quenching in the presence of thiol-containing compounds, which is attributed to the selective addition of thiols to the cyclobutene ring in the chemosensors. These are two representative examples of thiol chemosensors that cannot distinguish Cys/Hcy and GSH.

Liu group reported a pyronin B based fluorescent chemosensor (2) for the discrimination of Cys/Hcy and GSH using different emission channels [27]. Initially, free (2) is non-fluorescent due to the PET process from the methoxythiophenol group to the pyronin moiety. Upon treatment of (2) with Cys, an intramolecular rearrangement occurs followed by a substitution reaction, which leads to fluorescence enhancement at 546 nm. The chemosensor has been applied to imaging in live cells. Importantly, the use of the intramolecular rearrangement of Cys with chemosensors is a typical strategy for the design of fluorescent chemosensors to discriminate Cys/Hcy and GSH.

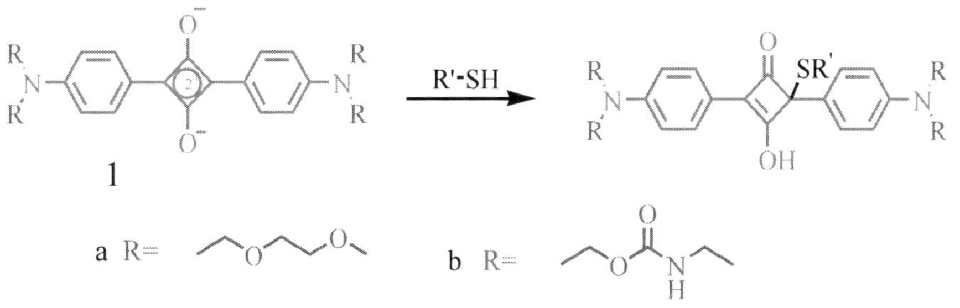

Scheme 1. Binding mechanism of **1** for cysteine.

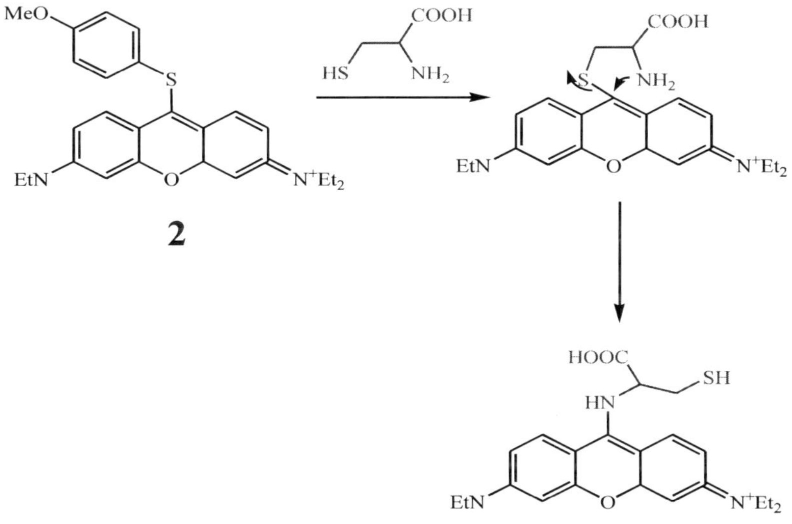

Scheme 2. Synthesis of compound (**2**) and it's binding with cysteine.

In 2019 Zhang et al. reported a benzothiazole-based near-infrared (NIR) ratiometric fluorescent probe (HBT-Cys) [28] for discriminating cysteine (Cys) from homocysteine (Hcy) and glutathione (GSH). The probe was designed by masking phenol group in the conjugated benzothiazole derivative with methacrylate group that could be selectively removed by Cys, and therefore an intramolecular charge transfer (ICT) fluorescence was switched on in the NIR region. In the absence of Cys, the probe exhibited a strong blue fluorescence emission at 431 nm, whereas a NIR fluorescence emission at 710 nm was significantly enhanced accompanied by a decrease of emission at 431 nm in the presence of Cys, allowing a ratiometric fluorescence detection of Cys. The fluorescence intensity ratio (I_{710nm}/I_{431nm}) showed a good linear relationship with Cys concentration of 1–40μM with the detection limit of 0.5μM. The sensing mechanism was explored based on MS experimental analysis and DFT theoretical calculation. Moreover, the fluorescent probe was successfully used for fluorescence bioimaging of Cys in living cells.

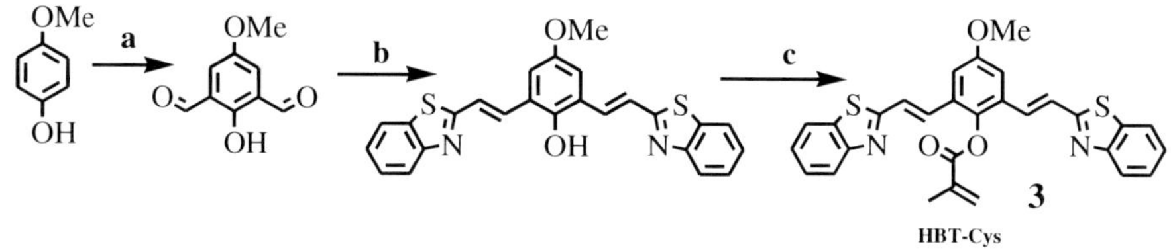

Scheme 3. Synthesis mechanism of HBT-Cys. Reagents and conditions: (a) hexamethylenetetramine, TFA, 110°C, 72 h; (b) 2-methylbenzothiazole, AcO$_2$, 145°C, 56 h; pyridine, 115°C, 2 h; water, 100°C, 6 h; and (c) methacryloyl chloride, acetone, K$_2$CO$_3$, r.t., 8 h.

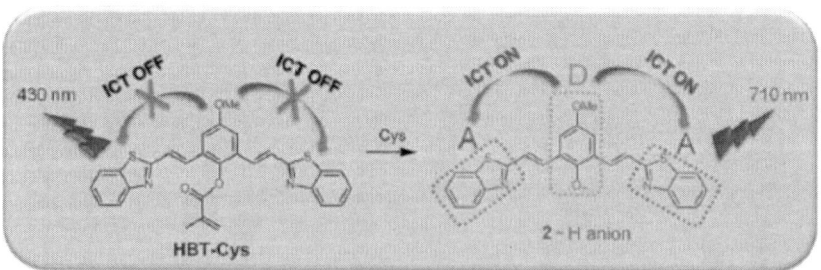

Figure 1. Sensing mechanism.

HBT-Cys (**3**) was synthesized by dissolving the compound 2,6-bis((E)-2-(benzo[d]thiazol-2-yl)vinyl)-4-methoxyphenol (0.1 mmol) in acetone (200 mL) and methacryloyl chloride (0.13 mmol in 5 mL acetone) was slowly added dropwise under stirring at 0°C in the presence of K$_2$CO$_3$ (2.0 mmol). Then the mixture was warmed to room temperature and stirred for another 8 h. The reaction mixture was filtered and the filtrate was concentrated in vacuum. The residue was dissolved in CH$_2$Cl$_2$ and then washed with water and the organic layer was dried with MgSO$_4$. After removing MgSO$_4$ by filtration, the crude product was obtained by evaporation under reduced pressure and then purified by column chromatography (silica gel, CH$_2$Cl$_2$/ethyl acetate = 30:1 v/v) to give HBT-Cys as yellow solid.

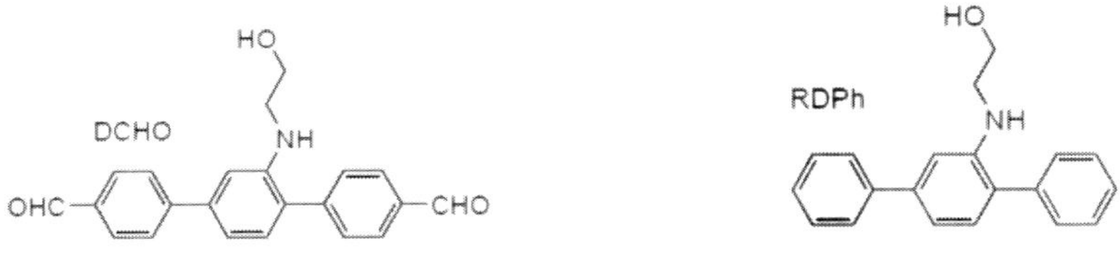

Figure 2. Chemical structure of DCHO (**4**) and the reference compound RDPh (**5**).

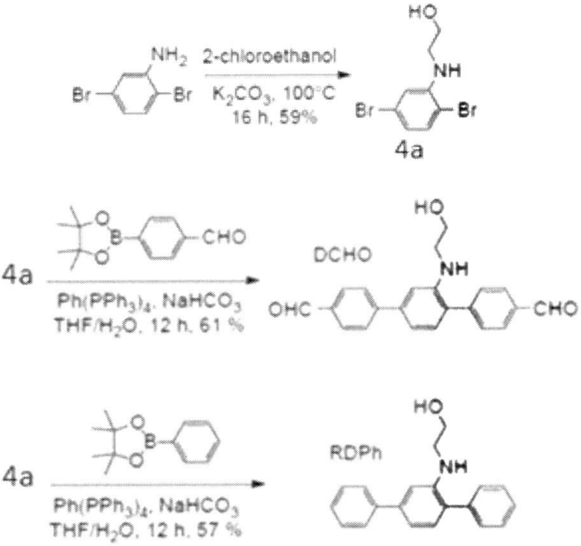

Scheme 4. Synthesis of conjugated compounds DCHO and RDPh.

In 2010 Y. Wang et al. reported an amphiphilic oligo p-phenylene derivative (DCHO) bearing electron-donating group [-NH(CH$_2$)$_2$OH] and electron withdrawing group (-CHO) for synthetic modification [29]. The sensing characteristics of this probe (DCHO) for cysteine (Cys) was studied in a mixture solution of DMSO–HEPES by UV–Vis and fluorescence spectra. ^{1}H-NMR, MALDI-TOF and UV–Vis titration experiments proved that thiazolidine and thiazinane derivatives were formed. The highly Cys-selective fluorescence hypochromic shift (>110 nm) can be observed due to the switching of intramolecular charge transfer, leading to potential fabrication of ratiometric fluorescent detection of Cys.

Compound DCHO and the reference compound RDPh were readily prepared via Suzuki cross-coupling reactions of N-hydroxyethyl-2,5-dibromoaniline and 4-formylbenzeneboronic acid pinacol ester or benzeneboronic acid pinacol ester, respectively, in the presence of Pd(PPh$_3$)$_4$ and NaHCO$_3$ in THF/H$_2$O solution. The hydroxyethyl group was introduced in DCHO to increase the hydrophile. The resulting molecules were characterized by FT-IR, ^{1}H-NMR, ^{13}C-NMR and mass spectroscopy.

N-(2-Hydroxyethyl)-2,5-dibromoaniline (compound **4a**) was synthesized by taking a mixture of 2,5-dibromoaniline (1.50 g, 6.0 mmol), 2- chloroethanol (36 ml) and K$_2$CO$_3$ (2.48 g, 18.0 mmol) stirred at 100^{0}C for 16 h under an atmosphere of nitrogen. The reaction was then cooled to ambient temperature, brine was added (100 ml) and the solution was extracted with CH$_2$Cl$_2$ (3 × 50 ml). The combined organic phase was collected and dried over Na$_2$SO$_4$, and filtered. The solvent was removed under reduced pressure and the crude product was purified by column chromatography (silica gel) with petroleum ether-CH$_2$Cl$_2$ (v/v, 10:3) as the eluent to give white solid **4a** (1.03 g, 59%).

An effective ratiometric fluorescent probe ethyl 2-(3-formyl-4-hydroxyphenyl)-4-methylthiazole-5-carboxylate (**6**) was reported by Risong Na et al. in 2016 [30] for the rapid and selective detection of cysteine in aqueous media. **6** was synthesized as a ratiometric fluorescent probe for the rapid and selective detection of Cys over glutathione (GSH) and other amino acids. The fluorescence intensity of the probe in the presence of Cys increased about 3-fold at a concentration of 20 equiv. of the probe, compared with that in the absence of these chemicals in aqueous media. The limits of detection of the fluorescent assay were 0.911 µM of Cys. ^{1}H-NMR and Mass Spectral analyses indicated that an excited-state intramolecular proton transfer is the mechanism of fluorescence sensing. This ratiometric probe is structurally simple and highly selective with the short reaction time of 2.5 h at ambient temperature. The results suggest that it has useful applications in analytical chemistry and diagnostics. They performed the UV-Vis absorption and fluorescence selectivity studies for the probe.

This probe undergoes an excited-state intramolecular proton transfer (ESIPT) process, which is confirmed by ^{1}H-NMR and MS spectra. The compound **6** at λ_{ex} 327 nm produced two emission peaks at ~426 nm and ~501 nm. Adding Cys to **6** enhanced the ESIPT process to shift the emission signal to a longer wavelength. As a result, the fluorescence intensity at λ_{em} 501 nm, which was attributed to the tautomer (T* emission) of the Cys - products, was enhanced strongly, and the normal isomer (N* emission) of the products caused the fluorescence intensity at 426 nm to increase slowly. The two emission peaks can be used for the ratiometric fluorescent measurement to determine the presence of the analytes more accurately with the minimization of the background signal. **6** was characterized and optimized as a fluorescent probe via studies of its UV-Vis absorption, fluorescence selectivity, reaction time of biothiols, quantification of Cys/Hcy, effect of pH on detection of Cys, as well as its detection mechanisms.

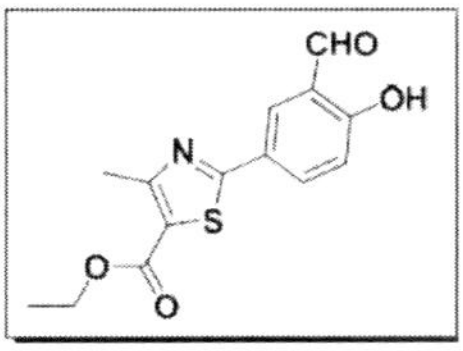

Figure 3. Structure of the fluorescent probe **6**.

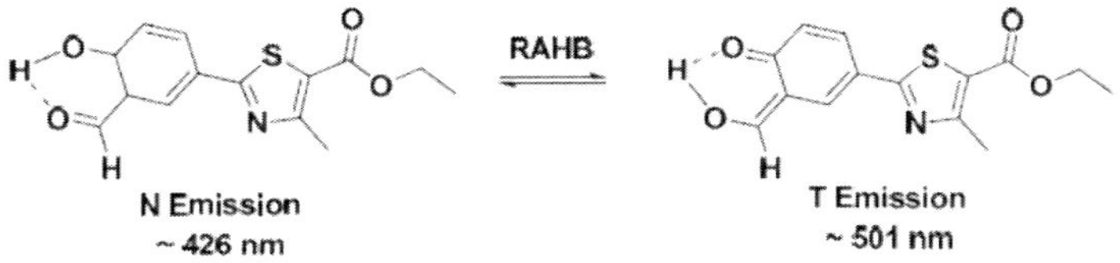

Scheme 5. Probable isomer of **6**.

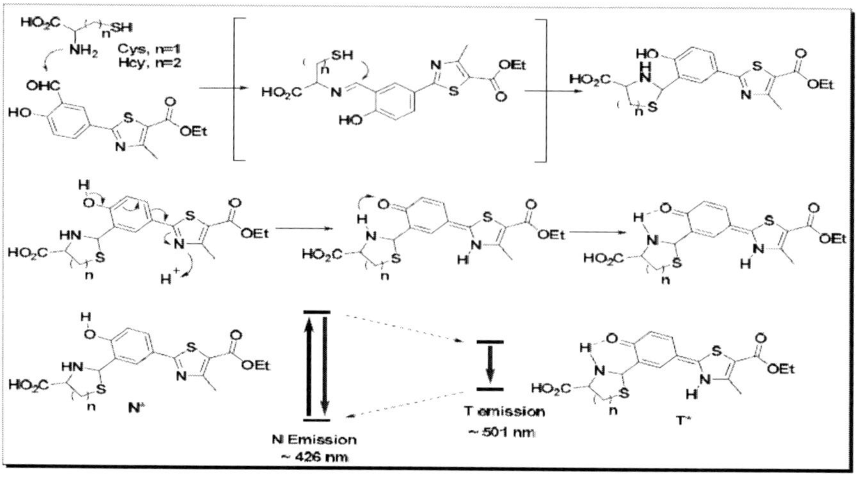

Scheme 6. Proposed reaction mechanism of **6** and Cys.

Chai et al. in 2018 reported a novel fluorescent sensor **8** from 8-aminoquinoline via intermediate **7,** for selective and sensitive detection of cysteine based on a complex between bi-8-carboxamidoquinoline derivative ligand (**8**) and Cu^{2+} [31]. The interaction of Cu^{2+} with the ligand causes a dramatic fluorescence quenching most likely due to its high affinity towards Cu^{2+} and a ligand–metal charge transfer (LMCT) process. The in situ generated **8**-Cu^{2+} complex was utilized as a chemosensing ensemble for cysteine. In the presence of cysteine, the fluorophore, **8**, was released from **8**-Cu^{2+} complex because of the strong affinity of cysteine to Cu^{2+} via the Cu–S bond, leading to the fluorescence recovery of the ligand. The proposed displacement mechanism was confirmed by the results of mass spectrometry (MS) study. Under optimized conditions, the recovered fluorescence intensity is linear with cysteine concentrations in the range 1×10^{-6} mol/L to 8×10^{-6} mol/L. The detection limit for cysteine is 1.92×10^{-7} mol/L. Furthermore, the established method showed a highly sensitive and selective response to cysteine among the 20 fundamental α-amino acids used as the building blocks of proteins, after Ni^{2+} was used as a masking agent to eliminate the interference of His. The sensor is applicable in monitoring cysteine in practical samples with good recovery rate.

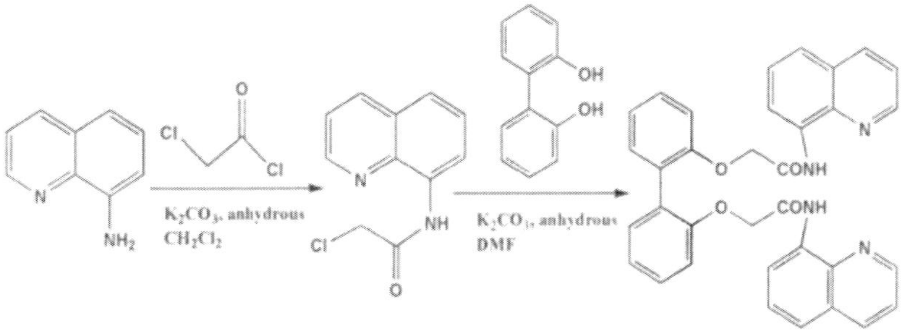

Scheme 7. Synthesis of **7** and ligand **8**.

The ligand **8** exhibited strong fluorescence, which was effectively quenched by complexation with Cu^{2+}. With the addition of Cys, significant fluorescence enhancement was observed due to the decomplexation of Cu^{2+} from the **8**-Cu^{2+} complex to release the fluorophore **8**. For the sake of a steadier as well as a better fluorescence response, several experimental conditions such as solvents, solvent composition, pH values, buffer solutions, incubation time and incubation temperature were optimized. In addition, the sensor exhibits excellent selectivity for Cys detection over other competing natural α-amino acids. As for the interference caused by His, Ni^{2+} was selected as a masking agent to eliminate this. Furthermore, the proposed method was applied successfully to the detection of Cys in honey samples with satisfactory results. The complex **8**-Cu^{2+} can be utilized as a sensitive and selective fluorescent sensor for Cys analysis over other naturally occurring α-amino acids, which could meet the selective requirements for practical applications.

Yuanyuan Li from Henan University, China in 2017 reported a ratiometric fluorescent chemosensor 2-(2'-hydroxy-3'-formyl-5'-methoxyphenyl) benzothiazole (**9**) for the detection of cysteine in aqueous solution at neutral pH [32]. In aqueous solution at neutral pH, **9** exhibited a ratiometric fluorescent response to Cys with a remarkable red to green shift in the emission wavelength. This fluorescence change was attributed to the cyclization reaction between the formyl group in **9** and the amino and sulfhydryl group in Cys in a stoichiometry of 1: 1 according to the proposed mechanism. At neutral pH, **9** displayed a significant fluorescence ratio signal enhancement with the addition of Cys. Furthermore, it also showed good selectivity towards Cys. The detection limit and linear range were 5.6 µmol/L and 0–100 µmol/L, respectively, which demonstrated that **9** could recognize relatively low concentrations of Cys and is a good candidate for applications in detecting Cys.

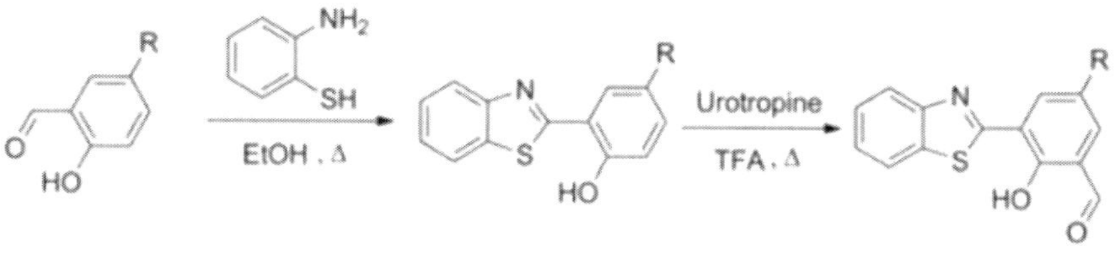

Scheme 8. Synthesis of compound **9**.

The compound **9a** was synthesized by dissolving 2-Hydroxy-4-methoxybenzaldehyde (4.56 g, 30 mmol) was dissolved in 30 ml EtOH, followed by the addition of 2-aminothiophenol (3.75 g, 30 mmol) and 5 ml 30% H_2O_2. The mixture was stirred at 80°C for 12 h to form a yellow precipitate. The product was then filtered and collected. After being dried under reduced pressure, 5.01 g of **9a** was obtained (yield 65%). 2-(2′-Hydroxy-3′-formyl-5′-methoxyphenyl) benzothiazole (**9**) was synthesized by mixing **9a** (2.57 g, 10 mmol) with 30 ml trifluoroacetic acid, then stirred and heated to 75°C until it dissolved. Urotropine (4.2 g, 30 mmol) was then added and the mixture was stirred at 80°C for 12 h. The solvent was removed by vacuum distillation to obtain a crude product. The crude product was purified on a silica-gel column using petroleum ether/ethyl acetate (5: 1 v/v) as the eluent. 0.77 g of **9** was obtained with a yield of 27%.

The reaction between **9** and Cys was first investigated using the time-dependent fluorescence and absorption spectra. The absorption peak at 467 nm decreased gradually and a new band at 359 nm appeared within 1 h after the addition of Cys (500 µmol/L) to **9** (10 µmol/L). Two isosbestic points at

Optical Sensors for Cysteine 97

395 and 332 nm could be observed, which suggested that new compound was formed. According to reports, the formyl group is an active site in the reaction of chemosensors with Cys. Thus, a proposed mechanism is shown in Scheme 9. After a cyclization reaction between the formyl group in **9** and the amino and sulfhydryl group in Cys, **9** became **9**-Cys along with a blue-shift in the fluorescence. To verify this proposed mechanism, a series of experiments were performed.

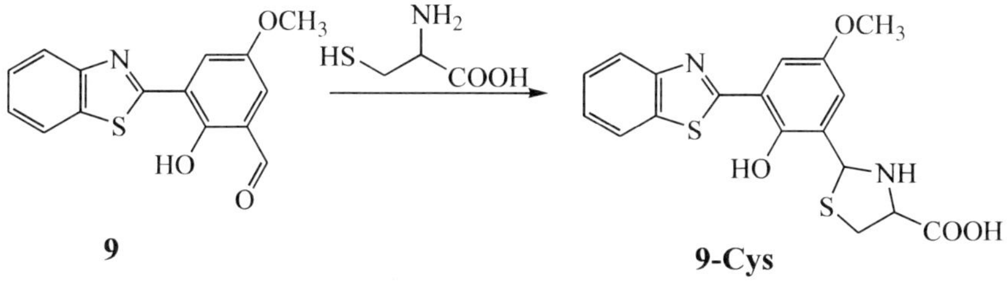

Scheme 9. Proposed mechanism for the reaction of **9** with Cys.

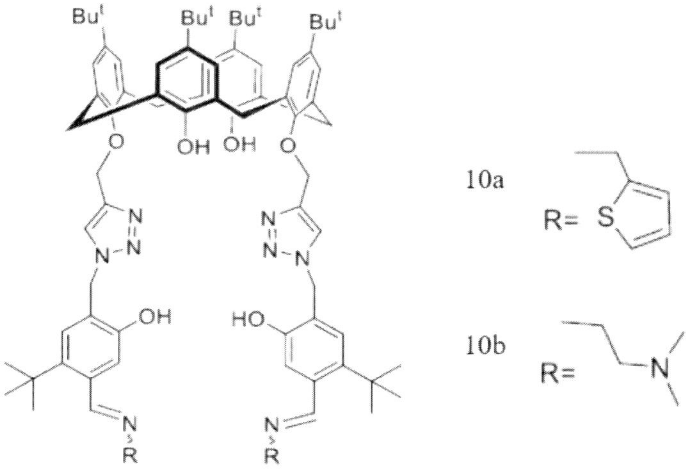

Figure 4. Structure of calyx [4] arene based probe **10a** and **10b**.

Rao et al. observed how fine-tuning of the receptor binding sites can lead to significant changes in their selectivity. For example, imine-based conjugate **10a** (Figure 4) [33] possessed weak fluorescence (λ_{em} = 454 nm,

λ_{ex} = 390 nm, 1:2 v/v HEPES buffer/EtOH) because of PET between the imine nitrogens and the two salicyl groups. Out of 12 mono- and divalent metal ions tested, **10a** exhibited a turn-on fluorescence response only upon treatment with Zn^{2+} ion. In this complex, the PET was blocked by the bound Zn^{2+}. The resulting complex served as a secondary chemosensor for the detection of thiol-containing mimics of metallothioneins (cysteine/dithiothreitol) [34]. Upon the addition of thiols, Zn^{2+} is displaced from the complex, resulting in restoration of the PET and the fluorescence attenuation. The potential of the **10a**-Zn^{2+} complex to recognize thiol-containing cysteine was also demonstrated in HeLa cells [35].

The same research group developed (N,N-dimethylamino)-ethylimino-bearing triazole-linked chemosensor **10b**, which exhibited a turn-on fluorescence response at 456 nm (λ_{ex} = 380 nm) upon complexation with Cd^{2+} (Figure 4). This **10b**-Cd^{2+} complex was further utilized for recognition of Cys among the 20 biologically important amino acids. In the case of Cys, the fluorescence of the complex was attenuated, and there was a color change from fluorescent blue to nonfluorescent. During the addition of Cys to a methanolic solution of **10b**-Cd^{2+}, a fluorescence enhancement was initially observed, followed by gradual quenching until saturation. This was attributed to the coordination of the thiol of Cys with Cd^{2+} followed by its removal from the coordination sphere of the host. The chemosensor **10b**-Cd^{2+} exhibited good reversibility and was further utilized for sensing of Cys in biological environments.

REFERENCES

[1] Liu, J., Sun, Y.-Q., Huo, Y., Zhang, H., Wang, L., Zhang, P., (2014). "Simultaneous fluorescence sensing of Cys and GSH from different emission channels." *J. Am. Chem. Soc.* 136:574–577.

[2] Lipton, S. A., Choi, Y. B., Takahashi, H., Zhang, D., Li, W., Godzik, A. (2002). "Cysteine regulation of protein function - as exemplified by NMDA-receptor modulation." *Trends Neurosci.* 25:474–480.

[3] Shao, J., Sun, H., Guo, H., Ji, S., Zhao, J., Wu, W. (2012). "A highly selective red-emitting FRET fluorescent molecular probe derived from BODIPY for the detection of cysteine and homocysteine: an experimental and theoretical study." *Chem. Sci. 3:1049–1061.*

[4] Qi, Y., Huang, Y., Li, B., Zeng, F., and Wu, S. (2018). "Real-time monitoring of endogenous cysteine levels in vivo by near-infrared turn-on fluorescent probe with large stokes shift." *Anal. Chem. 90: 1014–1020.*

[5] Shahrokhian, S. (2001). "Lead phthalocyanine as a selective carrier for preparation of a cysteine-selective electrode." *Anal. Chem.* 73:5972–5978.

[6] Giles, N. M., Watts, A. B., Giles, G. I., Fry, F. H., Littlechild, J. A., and Jacob, C. (2003). "Metal and redox modulation of cysteine protein function." *Chem. Biol.* 10, 677–693.

[7] Dai, X., Wu, Q. H., Wang, P. C., Tian, J., Xu, Y., Wang, S. Q., Miao, J. Y., Zhao, B. X. (2014) "A simple and effective coumarin-based fluorescent probe for cysteine." *Biosens. Bioelectron.* 59:35–39.

[8] Zhou, X., Jin, X., Sun, G., Wu, X. (2013). "A sensitive and selective fluorescent probe for cysteine based on a new response-assisted electrostatic attraction strategy: The role of spatial charge configuration." *Chemistry*, 19:7817–7824.

[9] Shahrokhian, S. (2001) "Lead phthalocyanine as a selective carrier for preparation of a cysteine-selective electrode." *Anal. Chem.* 73:5972–5978.

[10] Wang, W., Rusin, O., Xu, X., Kim, K. K., Escobedo, J. O., Fakayode, S. O., Fletcher, K. A., Lowry, M., Schowalter, C. M., Lawrence, C. M. (2005). "Detection of homocysteine and cysteine." *J. Am. Chem. Soc.* 2005, 127:15949–15958.

[11] Paulsen, C. E., Carroll, K. S. (2013). "Cysteine-mediated redox signaling: Chemistry, biology, and tools for discovery." *Chem. Rev.*, 113: 4633–4679.

[12] Refsum, H., Smith, A. D., Ueland, P. M., Nexo, E., Clarke, R., McPartlin, J., Johnston, C., Engbaek, F., Schneede, J., McPartlin, C.

(2004). "Facts and recommendations about total homocysteine determinations: An expert opinion." *Clin. Chem.*, 50:3–32.

[13] Nekrassova, O., Lawrence, N. S., Compton, R. G. (2003). "Analytical determination of homocysteine: A review." *Talanta*, 60:1085–1095.

[14] Peng, H., Chen, W., Cheng, Y., Hakuna, L., Strongin, R., Wang, B. (2012). "Thiol reactive probes and chemosensors." *Sensors* 2012, 12:15907–15946.

[15] Zheng, C., Pu, S., Liu, G., Chen, B., Dai, Y. (2013). "A highly selective colorimetric sensor for cysteine and homocysteine based on a new photochromic diarylethene." *Dyes Pigment.*, 98:280–285.

[16] Mei, J., Wang, Y., Tong, J., Wang, J., Qin, A., Sun, J. Z., Tang, B. Z. (2013). "Discriminatory detection of cysteine and homocysteine based on dialdehyde-functionalized aggregation-induced emission fluorophores." *Chem. A Eur. J.*, 19:613–620.

[17] Goswami, S., Manna, A., Paul, S., Das, A. K., Nandi, P. K., Maity, A. K., Saha, P. (2014). "A turn on esipt probe for rapid and ratiometric fluorogenic detection of homocysteine and cysteine in water with live cell-imaging." *Tetrahedron Lett.*, 55:490–494.

[18] Han, C., Yang, H., Chen, M., Su, Q., Feng, W., Li, F. (2015). "Mitochondria-targeted near-infrared fluorescent off-on probe for selective detection of cysteine in living cells and in vivo." *ACS Appl. Mater. Interfaces*, 7:27968–27975.

[19] Chen, Z., Sun, Q., Yao, Y., Fan, X., Zhang, W., and Qian, J. (2017). Highly sensitive detection of cysteine over glutathione and homocysteine: new insight into the Michael addition of mercapto group to maleimide. *Biosens. Bioelectron.* 91, 553–559.

[20] Li, Y., Liu, W., Zhang, P., Zhang, H., Wu, J., Ge, J., et al. (2017). A fluorescent probe for the efficient discrimination of Cys, Hcy and GSH based on different cascade reactions. *Biosens. Bioelectron.* 90, 117–124.

[21] Nawimanage, R. R., Prasai, B., Hettiarachchi, S. U., and McCarley, R. L. (2017). Cascade reaction-based, near-infrared multiphoton fluorescent probe for the selective detection of cysteine. *Anal. Chem.* 89, 6886–6892.

[22] Sheng, H., Hu, Y., Zhou, Y., Fan, S., Cao, Y., Zhao, X., et al. (2019). A highly selective ESIPT-based fluorescent probe with a large Stokes shift for the turn-on detection of cysteine and its application in living cells. *Dyes Pigm.* 160, 48–57.

[23] Valeur, B., Leray, I. (2000). *Coord. Chem. Rev.*, 205, 3–40. (b) Rurack, K. (2001). *Spectrochim. Acta Part A,* 57:2161–2195.

[24] Ward, M. D. (1997). *Chem. Soc. Rev.*, 26:365–375. (b) Armaroli, N., Accorsi, G., Felder, D., Nierengarten, J. F. (2002). Chem. Eur. J., 8:2314–2323. (c) Jo¨dicke, C. J., Lu¨thi, H. P. J. (2003). Am. Chem. Soc., 125:252–264. (d) Demeter, A., Be´rces, T., Zachariasse, K. A. (2001). *J. Phys. Chem. A*, 105:4611–4621.).

[25] de Silva, A. P., Gunaratne, H. Q. N., and Gunnlaugsson, T., *Tetrahedron Lett.*, 1998, 39, 5077–5080.

[26] Ros-Lis, J. V., Garcı´a, B., Jime´nez, D., Martı´nez-Ma´n˜ez, R., Sanceno´n, F., Soto, J., Gonzalvo, F. and Valldecabres, M. C., *J. Am. Chem. Soc.*, 2004, 126, 4064–4065.

[27] Liu, J., Sun, Y.-Q., Zhang, H., Huo, Y., Shi, Y. and Guo, W., *Chem. Sci.*, 2014, 5, 3183–3188.

[28] Zhang, X., Zhang, L., Ma, W. W., Zhou, Y., Lu, Z. N. and Xu, S. (2019). "A Near-Infrared Ratiometric Fluorescent Probe for Highly Selective Recognition and Bioimaging of cysteine." *Frontiers in Chemistry* 7:32.

[29] Wang, Y., Xiao, J., Wang, S., Yang, B. and Ba, X. (2010). "Tunable fluorescent sensing of cysteine and homocysteine by intramolecular charge transfer." *Supramolecular Chemistry* 22:380-386.

[30] Na, R., Zhu, M., Fan, S., Wang, Z., Wu, X., Tang, J., Liu, J., Wang, Y. and Hua, R. (2016). "A Simple and Effective Ratiometric Fluorescent Probe for the Selective Detection of Cysteine and Homocysteine in Aqueous Media." *Molecules* 21:1023.

[31] Chai, G., Liu, Q., Fei, Q., Zhang, J., Sun, X., Shan, H., Feng, G., Huan, Y. (2017). "A selective and sensitive fluorescent sensor for cysteine detection based on bi-8-carboxamidoquinoline derivative and Cu^{2+} complex." *Luminescence* 33:153-160.

[32] Li, Y. (2017). "A ratiometric fluorescent chemosensor for the detection of cysteine in aqueous solution at neutral pH." *Luminescence* 32:1385-1390.

[33] Pathak, R. K., Hinge, V. K., Mondal, M., Rao, C. P. (2011). "Triazolelinked-thiophene conjugate of calix [4] arene: its selective recognition of Zn^{2+} and as biomimetic model in supporting the events of the metal detoxification and oxidative stress involving metallothionein." *J. Org. Chem.*, 76:10039−10049.

[34] Chang, C. J., Jaworski, J., Nolan, E. M., Sheng, M., Lippard, S. J. (2004). "A tautomeric zinc sensor for ratiometric fluorescence imaging: Application to nitric oxide-induced release of intracellular zinc." *Proc. Natl. Acad. Sci. U. S. A.*, 101:1129−1134.

[35] Pathak, R. K., Tabbasum, K., Rai, A., Panda, D., Rao, C. P. (2012). "A Zn2+ specific triazole based calix [4] arene conjugate (L) as a fluorescence sensor for histidine and cysteine in HEPES buffer." milieu. *Analyst*, 137:4069−4075.

INDEX

#

12-tungstophosphoric acid, 58, 60, 82
^{13}C-NMR, 92
^{1}H-NMR, 92, 93
2,6-bis((E)-2-(benzo[d]thiazol-2-yl)vinyl)-4-methoxyphenol, 90

A

acetaldehyde, 7, 10, 21
acetaminophen, 9, 23, 24
acid, viii, ix, 1, 2, 5, 12, 13, 15, 16, 17, 18, 24, 25, 27, 29, 33, 36, 49, 50, 57, 58, 59, 61, 62, 79, 81, 82, 85, 92
adsorption, 12, 16, 27, 28, 62, 68, 69
alternative medicine, 20, 56
amino acids, vii, ix, 1, 4, 6, 10, 11, 12, 13, 14, 15, 16, 17, 18, 19, 20, 29, 35, 49, 66, 71, 85, 86, 87, 92, 94, 95, 98
amino groups, 15
anterior cingulate cortex, 50
antioxidant, viii, 2, 3, 7, 10, 14, 17, 18, 35, 36, 37, 45, 47, 50, 52, 86
anxiety, 34, 42, 44, 51, 52
autoantibodies, 51
autoimmune disease, 38, 54

B

benefits, viii, 2, 3, 4, 5, 7, 10, 19, 21, 45, 47
benzene-boronic, 92
bioavailability, 36
biocompatibility, 11, 12
biological processes, ix, 85
biological roles, 6
biological systems, 88
biomedical applications, 10, 11, 18, 25
biomolecules, 11, 12, 80, 83, 87
blood, 10, 21, 35, 49, 53, 78
blood pressure, 10
blood stream, 35
blood-brain barrier, 49
body composition, 8, 22

brain, viii, 5, 20, 31, 32, 35, 36, 37, 39, 45, 48, 49, 50, 51, 52, 54, 78
bronchitis, vii, viii, 2, 3
building blocks, vii, 1, 13, 94

C

cardiovascular disease, vii, viii, ix, 2, 3, 7, 9, 21, 78, 85
characterized, 61, 92, 93
chemical, 12, 15, 18, 26, 63, 64, 86
chemical characteristics, 86
chemical properties, 15, 18
chemical reactivity, 86
chemokines, 45
chemosensors, 87, 88, 97, 100
children, ix, 5, 7, 13, 20, 23, 24, 32, 34, 39, 42, 43, 85, 87
chronic obstructive pulmonary disease, vii, viii, 2, 3
clinical trials, 6, 7, 8, 9, 40, 41, 43, 46
colorimetry, 86, 87
comorbidity, 33, 38, 41, 44, 46
compounds, vii, 1, 8, 17, 20, 64, 88, 91
corrosion, viii, 2, 3, 15, 16, 17, 18, 29
cosmetic, viii, 2, 3, 13, 18
cysteine, v, vii, viii, ix, 1, 2, 3, 4, 5, 6, 7, 8, 10, 11, 12, 13, 14, 15, 16, 17, 18, 19, 20, 21, 22, 23, 24, 25, 27, 28, 29, 31, 35, 36, 49, 50, 54, 55, 57, 58, 59, 61, 62, 65, 66, 67, 68, 70, 71, 73, 74, 75, 76, 77, 78, 80, 83, 85, 86, 87, 88, 89, 92, 94, 95, 98, 99, 100, 101, 102
cystine, 13, 14, 21, 35, 49, 52
cytokines, 24, 36, 38, 45, 47
cytotoxicity, 80, 82

D

deficiency, ix, 3, 5, 49, 85, 86

detection, vii, ix, 27, 85, 86, 87, 88, 90, 92, 93, 94, 95, 98, 99, 100, 101, 102
detoxification, ix, 7, 50, 85, 87, 102
diabetes, vii, viii, 2, 3, 7, 8, 9, 21, 24
diabetic patients, 21
diffusion, 72, 73, 74, 75, 76, 77
discrimination, 88, 100
diseases, 6, 38, 53, 56, 58, 86, 87
disorder, vii, viii, 31, 32, 33, 34, 40, 48, 51, 52, 53, 54, 55, 56
distilled water, 59, 60, 61
drug addict, 48
drug addiction, 48
drug delivery, 11, 78
drug release, 75, 76, 79
drug therapy, 20
drugs, vii, 1, 5, 6, 34, 44, 45, 46, 47, 48, 75

E

elders, ix, 32, 34, 44, 47
emission, 87, 88, 89, 93, 95, 98, 100
environment, 15, 64
environmental effects, 87
environments, viii, 2, 3, 98
ether, 12, 92, 96
evidence, viii, 2, 6, 24, 34, 38, 39, 46, 47, 50
experimental condition, 59, 95

F

fibroblast growth factor, 8, 22
filtration, 91
fluorescence, 87, 88, 89, 92, 93, 94, 95, 96, 97, 98, 102
fluorimetry, 86
fluvoxamine, 40, 41, 43
food industry, 14, 18
formation, viii, 14, 17, 32, 35, 61

functionalization, vii, ix, 11, 57, 58, 60, 64, 68, 71, 74

G

gastrointestinal tract, 10, 14
geometry, ix, 58, 59, 74, 77
glutamate, viii, 31, 32, 33, 34, 35, 36, 37, 38, 41, 45, 46, 47, 48, 49, 50, 51, 52
glutamine, 51
glutathione, viii, 2, 5, 7, 8, 9, 10, 18, 20, 22, 32, 35, 38, 49, 58, 87, 89, 92, 100
gold nanoparticles, 27
growth, ix, 5, 13, 85, 87

H

health, vii, viii, 2, 3, 5, 6, 7, 18
health status, 6, 18
hearing loss, 9, 10
heart disease, 7
heating rate, 62
hematopoiesis, ix, 85
hemocompatibility, 10
homeostasis, 19, 35, 36, 49, 87
homocysteine, 4, 9, 24, 87, 89, 99, 100, 101
human, viii, 2, 6, 12, 18, 25, 87
human body, viii, 2, 6, 12, 18
hydroxyl, 12, 17, 36, 50
hypochromic, 92
hypothesis, 35, 38, 39, 40, 41, 45, 49
hysteresis loop, 68

I

in vitro, ix, 57, 58, 76, 78, 80, 82, 83
in vitro release, ix, 57, 58, 59, 62, 71, 73, 77, 80, 82
in vivo, 12, 78, 99, 100
individuals, viii, 32, 45, 47

inflammation, vii, 7, 24, 45, 47, 50, 78
inflammatory mediators, 45
inhibition, 16, 17, 18, 23, 29, 33, 48, 49
inhibitor, viii, 2, 3, 11, 15, 16, 17, 18, 29, 54

K

kinetic model, ix, 57, 59, 75, 76
kinetics, ix, 57, 58, 59, 75, 77, 80, 82, 83
kinetics and mechanism, 58, 59, 77, 79

L

lipid peroxidation, 37, 52
liquid chromatography, 87
liver, 7, 9, 24, 25, 35, 44, 78, 86, 87
liver damage, 86, 87
liver disease, 24
liver failure, 9, 24, 78

M

magnetic resonance spectroscopy, 50
MCM-41, v, vii, ix, 57, 58, 59, 60, 61, 63, 64, 65, 66, 67, 68, 69, 70, 71, 72, 73, 74, 75, 76, 77, 79, 80, 82, 83
MCM-48, v, vii, ix, 57, 58, 59, 60, 61, 63, 64, 65, 66, 67, 68, 69, 70, 71, 72, 73, 74, 75, 76, 77, 79, 80, 82, 83
meat, 2, 14
medicine, vii, viii, 2, 3, 5, 6, 18
metabolic disorders, ix, 3, 85
metabolism, ix, 5, 6, 21, 37, 45, 52, 85, 87
metals, viii, 2, 3, 15, 16, 17, 18
methamphetamine, 9, 22
molecules, 6, 10, 13, 46, 63, 65, 73, 76, 87, 92
multiple sclerosis, 38, 53

N

N-Acetylcisteine, v, 31, 32
nanoparticles, 12, 27, 80, 83
natural compound, 6, 45
natural killer cell, 22, 38
neurobiology, 33, 39, 41, 45, 46, 51
neurodegenerative diseases, 78
neurogenesis, 45
neuroimmune, viii, 31, 32, 39, 54
neuroleptics, 33, 47
neuronal cells, 37
neurons, 36, 38
neuroprotection, 45
neuropsychiatry, 51
neurotoxicity, 45, 86
neurotransmission, viii, 31, 37, 45, 49
neurotransmitters, 45, 51
nitrogen, 5, 15, 18, 35, 62, 92
nutraceutical, ix, 5, 32, 34, 37, 46, 47

O

obsessive-compulsive disorder, viii, 31, 32, 48, 50, 51, 52, 53, 54, 55
optical sensors, v, vii, ix, 85, 86
organic compounds, 15
organism, ix, 10, 85, 86
orthostatic hypotension, 44
osteoarthritis, vii, viii, 2, 3
oxidation, 8, 22, 36, 81, 82, 86
oxidative stress, 8, 9, 10, 22, 23, 36, 37, 39, 47, 49, 50, 77, 102
oxygen, 15, 18, 73

P

pathways, viii, 31, 34, 35, 37, 45, 47, 48
pharmaceuticals, vii, 1, 5, 17, 37

pharmacological treatment, ix, 32, 37, 44, 45, 46, 47
photorefractive keratectomy, 10, 20, 22
physicochemical properties, 6
placebo, 40, 41, 42, 43, 50, 54, 55
population, 6, 32, 34, 41, 44, 53, 56
probe, 62, 89, 92, 93, 97, 99, 100, 101
protein hydrolysates, 14
protein synthesis, ix, 85, 87
protein-protein interactions, 14
proteins, vii, 1, 2, 5, 11, 13, 14, 19, 58, 86, 94
psychiatric disorder, 32, 36, 46
psychiatric disorders, 36, 46
psychosocial interventions, 33, 34, 44, 47

Q

quality of life, 43, 44, 47
quantification, ix, 85, 93
quantum dot, 12

R

reactions, 19, 26, 92, 100
reactivity, 6, 11, 26, 39, 87
receptor, 45, 49, 52, 97, 98
recognition, 98, 102
recognize, 6, 87, 96, 98
recommendations, iv, 39, 46, 100
resistance, 16, 22, 36, 43
response, 9, 33, 36, 39, 40, 41, 42, 44, 46, 47, 50, 78, 94, 95, 98, 99
room temperature, 61, 90

S

selectivity, 83, 87, 93, 95, 96, 97
sensing, 27, 90, 92, 93, 98, 101
sensor, 94, 95, 100, 101, 102

side effects, 40, 43, 44, 45, 47
silica, 26, 58, 78, 79, 80, 83, 91, 92, 96
skin, ix, 2, 5, 12, 13, 28, 43, 85, 87
solution, 11, 17, 29, 60, 61, 62, 92, 95, 98, 102
structure, 2, 6, 11, 13, 58, 59, 63, 64, 65, 68, 71, 77, 78, 91
sulfur, vii, 14, 15, 16, 18, 20, 58, 86
sulfuric acid, 16, 29
supplementation, viii, 10, 20, 21, 22, 23, 32, 40, 41, 46, 47
supramolecules, 86
symptoms, 6, 36, 40, 42, 43, 45, 47, 53, 56
synaptic plasticity, 52
synthesis, viii, 2, 4, 7, 8, 10, 20, 22, 27, 60, 78, 79, 80, 81, 82, 83
synthesized, 6, 17, 18, 60, 90, 92, 96
synthetic, 11, 91
synthetic polymers, 11
systemic lupus erythematosus, 53

T

T lymphocytes, 38

temperature, ix, 16, 17, 18, 57, 59, 60, 62, 81, 92, 93, 95
therapeutic effect, 40
therapeutic use, vii, viii, 32
therapy, 7, 8, 21, 23, 34, 35, 42, 44, 55
treatment, vii, viii, 2, 3, 6, 7, 8, 9, 13, 14, 21, 22, 23, 24, 32, 33, 34, 35, 36, 37, 38, 39, 40, 41, 42, 43, 44, 45, 46, 47, 50, 51, 54, 55, 62, 86, 88, 98
trial, 7, 25, 40, 41, 42, 43, 54, 55
trichotillomania, 48, 53
tumor necrosis factor, 53

V

vascular prostheses, 10, 25
viscoelastic properties, 12

W

water, 59, 60, 61, 65, 75, 90, 91, 100